T0325346

Crystalline Solid State Physics

An interactive guide

Online at: https://doi.org/10.1088/978-0-7503-5217-8

Crystalline Solid State Physics

An interactive guide

Meng Lee Leek

School of Physical and Mathematical Sciences, Nanyang Technological University, Singapore, Singapore

IOP Publishing, Bristol, UK

© IOP Publishing Ltd 2023

All rights reserved. No part of this publication may be reproduced, stored in a retrieval system or transmitted in any form or by any means, electronic, mechanical, photocopying, recording or otherwise, without the prior permission of the publisher, or as expressly permitted by law or under terms agreed with the appropriate rights organization. Multiple copying is permitted in accordance with the terms of licences issued by the Copyright Licensing Agency, the Copyright Clearance Centre and other reproduction rights organizations.

Permission to make use of IOP Publishing content other than as set out above may be sought at permissions@ioppublishing.org.

Meng Lee Leek has asserted his right to be identified as the author of this work in accordance with sections 77 and 78 of the Copyright, Designs and Patents Act 1988.

Multimedia content is available for this book from https://doi.org/10.1088/978-0-7503-5217-8.

ISBN   978-0-7503-5217-8 (ebook)
ISBN   978-0-7503-5215-4 (print)
ISBN   978-0-7503-5218-5 (myPrint)
ISBN   978-0-7503-5216-1 (mobi)

DOI   10.1088/978-0-7503-5217-8

Version: 20231201

IOP ebooks

British Library Cataloguing-in-Publication Data: A catalogue record for this book is available from the British Library.

Published by IOP Publishing, wholly owned by The Institute of Physics, London

IOP Publishing, No.2 The Distillery, Glassfields, Avon Street, Bristol, BS2 0GR, UK

US Office: IOP Publishing, Inc., 190 North Independence Mall West, Suite 601, Philadelphia, PA 19106, USA

Contents

Preface

This book grew out of teaching the module PH3102 Condensed Matter Physics I at Nanyang Technological University, School of Physical and Mathematical Sciences, Physics and Applied Physics Division.

This module is the first module in which students encounter the application of quantum mechanics; in this case, it is applied to the properties of solids. There are many properties of solids that have important technological applications, and an understanding of these properties usually requires quantum mechanics. In a typical first course in solid-state physics, the elastic properties of the lattice (lattice vibrations) and the electronic properties are covered, as they lead to the technology of electronics.

During the course of teaching this module, I have noticed that the students encounter several hurdles. Here, I wish to highlight two concepts that students have problems with:

- Reciprocal space: the student perceives a solid to be a crystal with atoms in a regular arrangement in real space. However, it turns out that most of the important physics of the solid is in its reciprocal space instead! The reciprocal space is essentially the (discrete) momentum space of the solid, and it is related to the real space by a Fourier series. Thus, the student is shocked by the fact that many properties of the solid are explained in the context of the abstract reciprocal space (available at https://doi.org/10.1088/978-0-7503-5217-8).
- Approximations: in a typical basic quantum mechanics course, the student always encounters nice, analytical examples. However, in the case of a solid, the problem is obviously too complex to be analytically solvable, so approximations are inevitable. The student was not trained to be aware of approximations, let alone to know how to deploy approximations.

The purpose of writing this book is to help students achieve a better understanding of the foundations of solid-state physics. This book has two features to assist students with that, which they have been tested in my teaching:

- Animations: as mentioned above, students are daunted by the introduction of the reciprocal space and the usage of reciprocal space. My ebook has a number of animations to illustrate the construction of the reciprocal space from real space and the usages of the reciprocal space (available at https://doi.org/10.1088/978-0-7503-5217-8).
- Detailed calculations: in order that the students do not become bogged down by the mathematical details of a calculation and the physical approximations used in a calculation, my ebook provides very detailed steps for nearly every calculation. This feature also allows the student to spend time pondering over the physical consequences of a calculation rather than struggling with filling in the mathematical steps of the calculation.

The layout of this ebook is as follows: chapter 1 covers the basic terminology of crystal structures and how x-ray diffraction of the crystal planes led to measurements of crystal structures. The basic physics and chemistry of bonding are then discussed to explain the stability of crystals.

Chapter 2 covers the basic elastic properties of the crystal lattice. The atomic vibrations are treated quantum mechanically in order to derive and explain the lattice's heat capacity at low temperatures.

Chapter 3 covers the basic electronic properties of solids.

Chapter 4 covers the basic developments in harnessing the electronic properties of solids in actual electronic devices. First, intrinsic (or unmodified) semiconductor solids are discussed; then, extrinsic (or modified) semiconductor solids are discussed. Finally, this leads to a discussion of the simplest electronic device, called the 'diode', which is made by joining two pieces of extrinsic semiconductor together.

Chapter 5 covers another important quantum mechanical property of solids: magnetism. Atomic magnetism is discussed first, which comes in the forms of diamagnetism or paramagnetism. A basic discussion of collective magnetism then completes the ebook. Collective magnetism exists due to long-range quantum mechanical interactions between magnetic moments.

My hope is that after reading this ebook, the student will become quite independent and will be able to dig deeper into the concepts discussed in this book or learn about other properties of solids from other sources.

Finally, any errors found in this ebook are inherently my responsibility. If you find any errors, please email me at: mlleek@ntu.edu.sg.

About the author

Meng Lee Leek

 Meng Lee Leek obtained his BSc (1st class honours), MSc, and PhD degrees from the National University of Singapore (NUS). Since 2012, he has been teaching at Nanyang Technological University of Singapore (NTU), School of Physical and Mathematical Sciences (SPMS), Physics and Applied Physics Division. From 2020, he has held the position of senior lecturer in physics, and from 2023 he has also held the Assistant Chair for Outreach. He was awarded the Nanyang Education Award (school) in 2017 and has been awarded the SPMS Teaching Excellence award three times. His research interests include theoretical high-energy physics and theoretical condensed matter. He has supervised numerous student projects (FYP, URECA, and Odyssey) and he enjoys showing students the mind-boggling world of theoretical physics.

IOP Publishing

Crystalline Solid State Physics
An interactive guide
Meng Lee Leek

Chapter 1

Crystal fundamentals

'Condensed matter' refers to matter in the solid or liquid state. In this book, we focus only on the study of solids, which is known as solid-state physics. The simplest solids are those with more symmetries—these are called 'crystalline solids' or 'crystals'.

We start off our study of crystalline solids by covering three fundamental aspects: (1) crystallography (section 1.1), (2) x-ray diffraction (section 1.2), and (3) binding in solids (section 1.3).

In the section on crystallography, we cover basic definitions and 2D and 3D examples to familiarise the reader with the basic terminology used in crystallography.

X-ray diffraction is the standard way to probe the internal structures of solids. The important concepts of reciprocal space and Brillouin zones will be introduced.

The section on binding in solids considers how solids are held together in a stable manner.

1.1 Crystal structures

1.1.1 Eleven basic definitions

In the study of solids, it is convenient and useful to assume that solids are made up of infinite, perfectly regular crystal structures. Obviously, this is not true in real life. In real life, crystals are not infinite and are not perfectly regular throughout.

Under these two simplifying assumptions, it is useful to learn a series of terms that describe infinite and perfect crystals. The following definitions are the important terms that we must learn in the study of solids.

1.1.1.1 Definition 1: Bravais lattice

A Bravais lattice is a set of points such that the environment around every point is the same. This set of points can be mathematical, i.e. the points are imagined and there may actually be no atoms there. There are two points to note:

doi:10.1088/978-0-7503-5217-8ch1

© IOP Publishing Ltd 2023

1. There is a set of translation vectors $\vec{R}$ that links every point to every other point.[1]
2. Due to symmetry operations (see section 1.5), there are only five types of Bravais lattices in 2D and 14 types in 3D. We will examine these later.

1.1.1.2 Definition 2: basis
A basis is a structural unit that is repeated at every Bravais lattice point. A basis can be made out of one atom, a few atoms, or something more complicated.

1.1.1.3 Definition 3: crystal
A crystal is a Bravais lattice with the same basis at every point. See figure 1.1.

Practically speaking, we are first faced with a crystal. We can check the environment around every atom in the crystal to see whether it is a Bravais lattice. If the environment around every atom is not the same, then we spend a few moments in thought to unravel its underlying Bravais lattice and its basis.

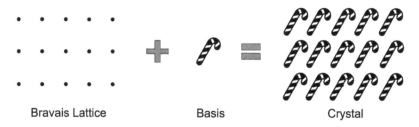

| Bravais Lattice | Basis | Crystal |

Figure 1.1. A crystal is a Bravais lattice that is 'decorated' with the same basis at every Bravais lattice point. Animation available at https://doi.org/10.1088/978-0-7503-5217-8.

1.1.1.4 Definition 4: primitive lattice translation vectors
The primitive lattice translation vectors are a set of (3D) vectors $\vec{a}_1$, $\vec{a}_2$, and $\vec{a}_3$ such that the translation vector $\vec{R} = n_1\vec{a}_1 + n_2\vec{a}_2 + n_3\vec{a}_3$ can reach any point in the lattice. Here, n_1, n_2, and n_3 are integers which can be positive or negative (figure 1.2).

1.1.1.5 Definition 5: primitive unit cell
A primitive unit cell is a volume of space that, when translated with all the possible $\vec{R}$ vectors, fills all the space without overlap or gaps. Note that there are infinite ways to define a primitive unit cell. See figure 1.3 for some examples

The following properties should be noted:
1. It should be clear that a primitive unit cell contains only one Bravais lattice point (which is not necessarily one atom!).
2. All choices of primitive unit cell have the same volume. In 3D, the volume[2] of the primitive unit cell is given by $V_C = |\vec{a}_1 \cdot (\vec{a}_2 \times \vec{a}_3)|$. In 2D, the volume is called the area, and the area is given by $V_C = |\vec{a}_1 \times \vec{a}_2|$.

---

[1] This set of translation vectors has infinitely many members because there are infinitely many points.
[2] Recall that this is how you find the volume of a parallelepiped.

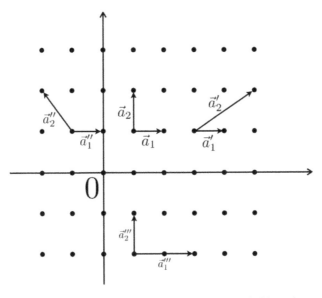

Figure 1.2. Can you tell why the sets $\vec{a}_1$, $\vec{a}_2$ and $\vec{a}'_1$, $\vec{a}'_2$ and $\vec{a}''_1$, $\vec{a}''_2$ are primitive, whereas the set $\vec{a}'''_1$, $\vec{a}'''_2$ is not primitive? Hint: think about whether you can reach any point in the lattice using suitable integer combinations for $\vec{R}$.

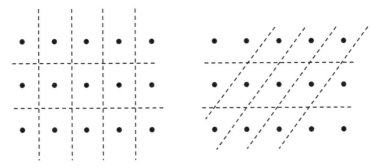

Figure 1.3. Two ways to choose primitive unit cells in a 2D Bravais lattice.

1.1.1.6 Definition 6: Wigner–Seitz primitive cell

The Wigner–Seitz primitive cell is a special choice of a primitive cell. It is a volume of space chosen such that the space in the cell is closer to one point than to any other point; this cell exhibits the full symmetry of the Bravais lattice.

Right now, this Wigner–Seitz choice of the primitive cell does not look useful or important. We discuss a very important use of the Wigner–Seitz cell in reciprocal space later in section 1.2.2.1.

The steps used to construct a Wigner–Seitz cell are:

Step 1:

Choose a point as the origin. Draw lines and connect the origin point to all the nearest neighbouring (nn) points (figure 1.4).

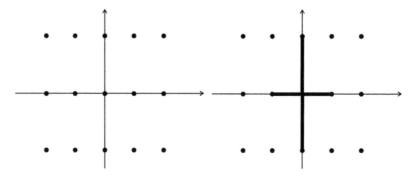

Figure 1.4. Two-dimensional example. LEFT: in step one, the centre point is chosen as the origin. RIGHT: there are four nearest neighbouring points, so four (thick) connecting lines are drawn. Animation available at https://doi.org/10.1088/978-0-7503-5217-8.

Step 2:

Mark the midpoints on all the connecting lines. Draw planes/lines that perpendicularly bisect the connecting lines (figure 1.5).

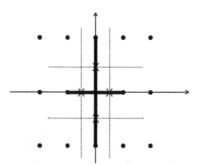

Figure 1.5. Two-dimensional example. In step two, the midpoints of the connecting lines are marked and four perpendicular bisector lines (planes in 3D) are drawn (in dotted lines). Animation available at https://doi.org/10.1088/978-0-7503-5217-8.

Step 3:

The smallest volume enclosed by the planes is the Wigner–Seitz primitive cell (figure 1.6).

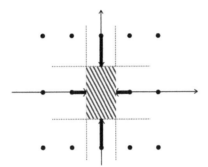

Figure 1.6. Two-dimensional example: The hatched area (volume in 3D) is the Wigner–Seitz primitive cell. Animation available at https://doi.org/10.1088/978-0-7503-5217-8.

1.1.1.7 Definition 7: conventional unit cell

A conventional unit cell is a volume of space that is usually bigger than a primitive unit cell and is chosen to explicitly exhibit certain symmetries of the crystal. Note that a conventional unit cell can cover all the space without an overlap or gaps by translation that uses a subset (not the full set) of $\vec{R}$ vectors. Choosing conventional unit cells makes discussions easier because in certain applications, it is not necessary to utilise the full symmetry of the solid.

1.1.1.8 Definition 8: The Miller index notation for planes

The Miller indices for planes in the (3D) crystal are denoted by three integers in round brackets (hkl).

The steps used to determine the Miller indices:

1. Find the intercepts between the plane and the axes formed by the primitive translation vectors. Note that in some cases, the plane and an axis may be parallel and extend to infinity. We then say that the intercept is at infinity.
2. Take the reciprocals of the numerical values of the intercepts.
3. Reduce them to integers in the simplest ratio and write them in round brackets. Negative integers are denoted by an overbar (figures 1.7 and 1.8).

We now turn to an example:

Step 1: the intercepts are at $3\vec{a}_1$, $1\vec{a}_2$, $2\vec{a}_3$

Step 2: the reciprocals are $\frac{1}{3}$, $\frac{1}{1}$, $\frac{1}{2}$

Step 3: multiply by six to get the integers 2, 6, 3, which are in the simplest ratio —this is the plane (263). Done.

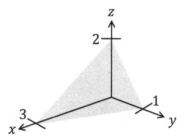

Figure 1.7. A three-dimensional example illustrating the Miller indices of a plane. Animation available at https://doi.org/10.1088/978-0-7503-5217-8.

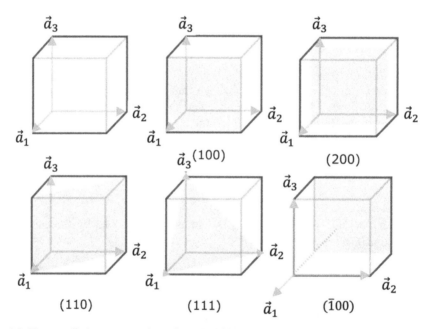

Figure 1.8. Here, we display common planes for a 3D cubic structure. The top-left diagram merely illustrates a cube for reference. Can you work out the Miller indices of the five planes shown?

Important sidetrack: reciprocal space (short story): It turns out that planes in a crystal are very important for x-ray studies, lattice waves, and electron waves in solids.

- We want to invent an abstract space to make it easier to keep track of planes in a crystal. This space is called reciprocal space. Points in this space represent (hkl) planes in real space.

- This sidetrack notes some of the basic properties of reciprocal space; the full story is given later (section 1.2.2).
- Just as we have the real space translation vector $\vec{R} = n_1\vec{a}_1 + n_2\vec{a}_2 + n_3\vec{a}_3$, so we define the reciprocal space translation vector as $\vec{G} = h\vec{b}_1 + k\vec{b}_2 + l\vec{b}_3$ so that indeed each point (in reciprocal space) represents an (*hkl*) plane in real space.
- This means that the primitive translation vectors in reciprocal space are $\vec{b}_1$, $\vec{b}_2$, and $\vec{b}_3$. They are defined by choosing $\vec{G}$ to be proportional to the normal vector of the plane (in real space). The choice that works (as we shall see in the next point) is

$$\vec{b}_1 = 2\pi\frac{\vec{a}_2 \times \vec{a}_3}{|\vec{a}_1 \cdot \vec{a}_2 \times \vec{a}_3|}, \quad \vec{b}_2 = 2\pi\frac{\vec{a}_3 \times \vec{a}_1}{|\vec{a}_1 \cdot \vec{a}_2 \times \vec{a}_3|}, \quad \vec{b}_3 = 2\pi\frac{\vec{a}_1 \times \vec{a}_2}{|\vec{a}_1 \cdot \vec{a}_2 \times \vec{a}_3|} \quad (1.1)$$

with $\vec{a}_i \cdot \vec{b}_j = 2\pi\delta_{ij}$.

- We shall show that $\vec{G} = h\vec{b}_1 + k\vec{b}_2 + l\vec{b}_3$ is indeed perpendicular to the corresponding (*hkl*) plane (in real space).
  - Assume the plane intercepts the axes at $u\vec{a}_1$, $v\vec{a}_2$, and $w\vec{a}_3$. The reciprocals are then $\frac{1}{u}$, $\frac{1}{v}$, and $\frac{1}{w}$.
  - We multiply by a (lowest) common multiple p to get h, k, l; i.e. $h = \frac{p}{u}$, $k = \frac{p}{v}$, and $l = \frac{p}{w}$.
  - A possible vector in the plane is $v\vec{a}_2 - u\vec{a}_1$.
  - We take the dot product between $\vec{G}$ and the vector in the plane: $\vec{G} \cdot (v\vec{a}_2 - u\vec{a}_1) = 2\pi(kv - hu) = \frac{2\pi p}{v}v - \frac{2\pi p}{u}u = 0$, thus proving $\vec{G}$ is normal to the plane.

The shortest distance between two planes in the family: Next, we want an expression for the shortest distance d between two parallel (*hkl*) planes which have the same Miller indices. You may wonder how parallel planes can be stacked with no space in between them. That is true for ordinary space, but in real space, planes only exist if they contain Bravais lattice points!

- We already checked that $\vec{G}$ is perpendicular to the plane, so $\frac{\vec{G}}{|\vec{G}|}$ is the unit normal vector to the plane.
- Looking at the plane that passes through the origin and the immediately adjacent plane, the distance d is thus obtained by the dot product between one intercept vector and the unit normal vector.
- See figure 1.9 to convince yourself that the intercept vectors of the immediately adjacent plane are $\frac{1}{h}\vec{a}_1$, $\frac{1}{k}\vec{a}_2$, and $\frac{1}{l}\vec{a}_3$. Therefore, when calculating the dot product between any of the intercept vectors, for example, $\frac{1}{h}\vec{a}_1$ and the unit normal $\frac{\vec{G}}{|\vec{G}|}$:

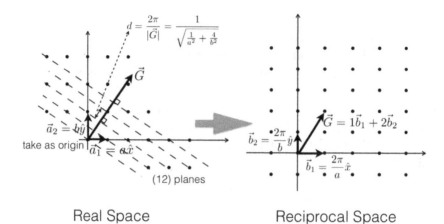

Real Space   Reciprocal Space

Figure 1.9. Two-dimensional examples of the reciprocal lattice and the distance between planes. The 'sequence of construction' should be like this: start with the real space primitive vectors $\vec{a}_1 = a\hat{x}$ and $\vec{a}_2 = b\hat{y}$; the reciprocal space primitive vectors are then $\vec{b}_1 = \frac{2\pi}{a}\hat{x}$ and $\vec{b}_2 = \frac{2\pi}{b}\hat{y}$. The primitive vectors $\vec{b}_1$ and $\vec{b}_2$ are constructed using $\vec{a}_i \cdot \vec{b}_j = 2\pi\delta_{ij}$. In reciprocal space, the generic translation vector is $\vec{G} = h\vec{b}_1 + k\vec{b}_2$. Take the (12) plane in real space as example; then $\vec{G} = 1\vec{b}_1 + 2\vec{b}_2$ (in reciprocal space) and the shortest distance between (12) planes (in real space) is $d = \frac{2\pi}{|\frac{2\pi}{a}\hat{x} + 2\frac{2\pi}{b}\hat{y}|} = \frac{1}{\sqrt{\frac{1}{a^2} + \frac{4}{b^2}}}$. Animation available at https://doi.org/10.1088/978-0-7503-5217-8.

$$d = \frac{1}{h}\vec{a}_1 \cdot \frac{\vec{G}}{|\vec{G}|}$$

| you can use the other intercept vectors $\frac{1}{k}\vec{a}_2$ or $\frac{1}{l}\vec{a}_3$, (1.2)

| and you will get the same answer at the end;

| recall $\vec{G} = h\vec{b}_1 + k\vec{b}_2 + l\vec{b}_3$

$$d = \frac{1}{h}\vec{a}_1 \cdot \frac{h\vec{b}_1 + k\vec{b}_2 + l\vec{b}_3}{|\vec{G}|}$$ (1.3)

| use $\vec{a}_i \cdot \vec{b}_j = 2\pi\delta_{ij}$

$$\boxed{d = \frac{2\pi}{|\vec{G}|}}$$ (1.4)

1.1.1.9 Definition 9: The Miller index notation for equivalent planes

Recall that the round bracket notation (hkl) labels a set of parallel planes. Following Miller, we shall use curly brackets to denote a family of planes $\{hkl\}$ that are related by symmetries (see section 1.5).

We use a 3D cube as an example.

family:$\{100\}$ = contains planes: (100), (010), (001), ($\bar{1}$00), (0$\bar{1}$0), (00$\bar{1}$)

family:$\{111\}$ = contains planes: (111), ($\bar{1}$11), (1$\bar{1}$1), (11$\bar{1}$), ($\bar{1}\bar{1}\bar{1}$), ($\bar{1}\bar{1}$1), ($\bar{1}$1$\bar{1}$), (1$\bar{1}\bar{1}$)

1.1.1.10 Definition 10: The Miller index notation for directions
Directions are denoted by square brackets.[3] These are simply direction vectors expressed in integer form. For example, if the direction vector is $1\vec{a}_1 + 2\vec{a}_2 + 3\vec{a}_3$, then the Miller index for the direction vector is [123]. See figure 1.10 for more examples.

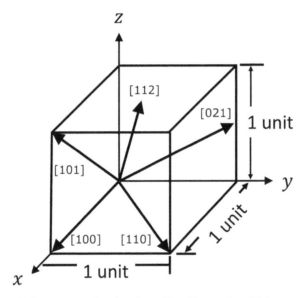

Figure 1.10. Here, we display common directions for a 3D cubic structure. Make sure you understand how each direction is obtained.

1.1.1.11 Definition 11: The Miller index notation for equivalent directions
Recall that the square bracket notation [*uvw*] labels a direction vector. Following Miller, we shall use angle brackets to denote a family of directions ⟨*uvw*⟩ that are related by symmetries (see section 1.5).

We use a 3D cube as an example.

family:⟨100⟩ = contains directions: [100], [010], [001], [$\bar{1}$00], [0$\bar{1}$0], [00$\bar{1}$]

---

[3] In cubic crystals, the direction [*hkl*] is perpendicular to the plane (*hkl*). This is a coincidence that happens because the primitive translation vectors are perpendicular to each other.

1.1.2 Concepts and examples illustrated by 2D crystals

Recall that a crystal is a Bravais lattice 'decorated' with the same basis at every point. The basis can be a single atom or a complicated molecule. Under the symmetry operations in 2D (see section 1.5), there are only five types of Bravais lattices possible (figure 1.11).

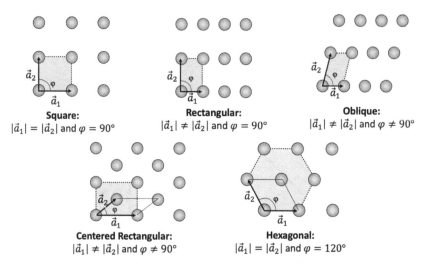

Figure 1.11. The five types of Bravais lattices in 2D. The translation vectors are all of the same form (as they should be): $\vec{R} = n_1\vec{a}_1 + n_2\vec{a}_2$. The primitive vectors $\vec{a}_1$, $\vec{a}_2$ are different across the five types.

1.1.2.1 The favourite 2D non-Bravais lattice example: the honeycomb lattice
- Consider the 2D honeycomb lattice below, which is found in popular materials such as graphene in which each lattice point is a carbon atom. Can you explain why the honeycomb lattice is not a Bravais lattice?
- Hint: look at the surroundings at each point—do the surroundings look the same? How many 'types' of surroundings are there? (figure 1.12)

- It turns out that two types of 'surroundings' can be found. So this is a two-atom-basis non-Bravais lattice.
  - Perspective 1: The underlying Bravais is a hexagonal lattice, and we can write two useful vectors (see figure 1.13) in Cartesian coordinates: $\vec{a}_1 = \frac{a\sqrt{3}}{2}\hat{x} + \frac{a}{2}\hat{y}$ and $\vec{a}_2 = \frac{a\sqrt{3}}{2}\hat{x} - \frac{a}{2}\hat{y}$. The two atoms are located at displacement vectors: $\vec{v}_1 = \frac{a}{2}\hat{y}$ and $\vec{v}_2 = -\frac{a}{2}\hat{y}$ from the hexagonal lattice points.

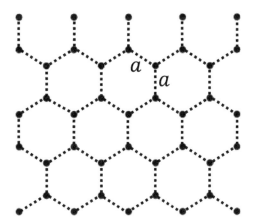

Figure 1.12. A two-dimensional honeycomb lattice. The dots are the carbon atoms and the dotted lines are the covalent bonds.

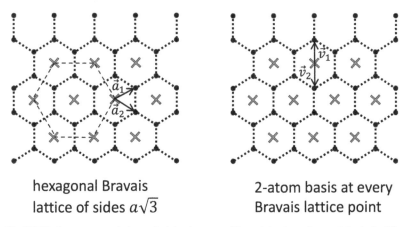

hexagonal Bravais
lattice of sides $a\sqrt{3}$

2-atom basis at every
Bravais lattice point

Figure 1.13. LEFT: the crosses mark the underlying hexagonal Bravais lattice points and the dashed lines serve as a guide to show the hexagonal Bravais lattice. RIGHT: the basis vectors $\vec{v}_1$ and $\vec{v}_2$ illustrate the locations of the carbon atoms relative to the hexagonal Bravais lattice points. Animation available at https://doi.org/10.1088/978-0-7503-5217-8.

    – Perspective 2: The underlying Bravais is an oblique lattice. See the exercise on how to write out the displacement vectors of the two atoms with respect to the oblique lattice points (figure 1.14).

1.1.3 Concepts and examples illustrated by 3D crystals

1.1.3.1 Seven crystal systems

Based on symmetry considerations (see section 1.5), there are more ways to 'play' with the angles and the lengths; thus, we get seven crystal systems in 3D (figure 1.15).

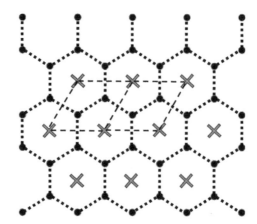

Figure 1.14. The underlying oblique Bravais lattice (indicated by dashed lines) for the honeycomb structure. Animation available at https://doi.org/10.1088/978-0-7503-5217-8.

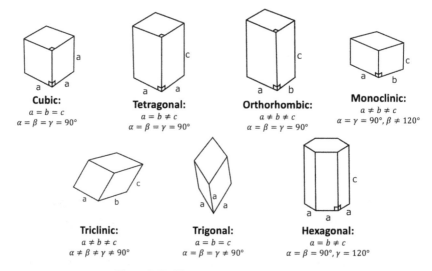

Figure 1.15. The seven crystal systems in 3D.

1.1.3.2 Fourteen types of Bravais lattices in 3D

The seven crystal systems are not yet Bravais lattices because in 3D, there are actually four ways to place the 'dots' (called 'centring') in these seven systems. This gives a total of 14 Bravais lattices (figure 1.16).

1.1.3.3 Simple 3D Bravais lattices: cubic Bravais lattices

Let us go through these simple but important 3D cubic structures.

Simple cubic (SC): the lattice has one point at each corner of the cube.

- The conventional cell is the same as the primitive cell.
- The primitive translation vectors are $\vec{a}_1 = a\hat{x}$, $\vec{a}_2 = a\hat{y}$, and $\vec{a}_3 = a\hat{z}$. See figure 1.17.

Lattice centring type:	Description:
Primitive (symbol P)	Lattice points at the corners only
Base-centred (symbols A (bc faces), B (ac faces), and C (ab faces))	Lattice points at the corners and at the centres of specific faces
Face-centred (Symbol F)	Lattice points at the corners and at the centres of all faces
Body-centred (Symbol I)	Lattice points at the corners and at the centre of the unit cell body

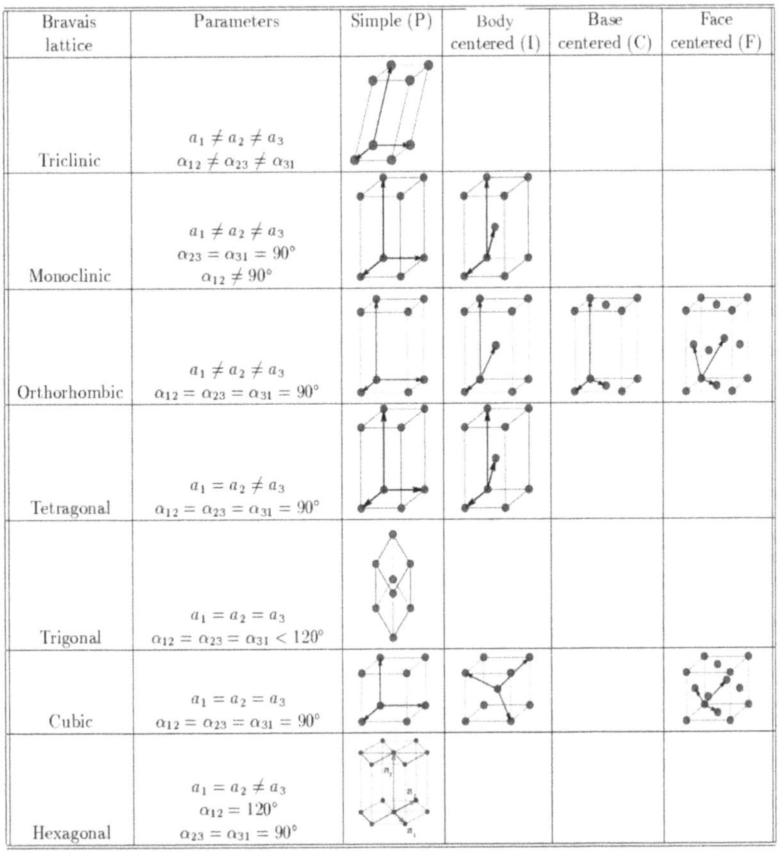

Bravais lattice	Parameters	Simple (P)	Body centered (I)	Base centered (C)	Face centered (F)
Triclinic	$a_1 \neq a_2 \neq a_3$ $\alpha_{12} \neq \alpha_{23} \neq \alpha_{31}$				
Monoclinic	$a_1 \neq a_2 \neq a_3$ $\alpha_{23} = \alpha_{31} = 90°$ $\alpha_{12} \neq 90°$				
Orthorhombic	$a_1 \neq a_2 \neq a_3$ $\alpha_{12} = \alpha_{23} = \alpha_{31} = 90°$				
Tetragonal	$a_1 = a_2 \neq a_3$ $\alpha_{12} = \alpha_{23} = \alpha_{31} = 90°$				
Trigonal	$a_1 = a_2 = a_3$ $\alpha_{12} = \alpha_{23} = \alpha_{31} < 120°$				
Cubic	$a_1 = a_2 = a_3$ $\alpha_{12} = \alpha_{23} = \alpha_{31} = 90°$				
Hexagonal	$a_1 = a_2 \neq a_3$ $\alpha_{12} = 120°$ $\alpha_{23} = \alpha_{31} = 90°$				

Figure 1.16. The 14 Bravais lattices in 3D. Reprinted from [1] with permission from Blackwell Publishing.

- Note that each point at the corner is shared by eight cells, so each corner point 'contributes' $\frac{1}{8}$ of a point to each cell. There are eight corners; thus, in total, SC has one point per cell.

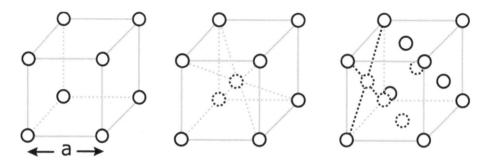

Figure 1.17. The three types of cubic Bravais lattices. All three cells have the lattice constant a. LEFT: Simple cubic (SC). MIDDLE: Body-centred cubic (BCC). RIGHT: Face-centred cubic (FCC).

- Under normal conditions, the only real-life example of the SC structure is the alpha phase of polonium. The reason that SC does not occur frequently in nature is because it does not pack the atoms efficiently (see section 1.1.3.5).

Body-centred cubic (BCC): The lattice has one point at each corner of the cube and one point at the centre of the cube.
- The conventional cell is not the primitive cell. The conventional cell has twice the volume of the primitive cell.
- Taking one of the corner points as the origin, the primitive translation vectors are $\vec{a}_1 = \frac{a}{2}(-\hat{x} + \hat{y} + \hat{z})$, $\vec{a}_2 = \frac{a}{2}(\hat{x} - \hat{y} + \hat{z})$, and $\vec{a}_3 = \frac{a}{2}(\hat{x} + \hat{y} - \hat{z})$. So the three primitive vectors point to three body-centred points. See figure 1.18.

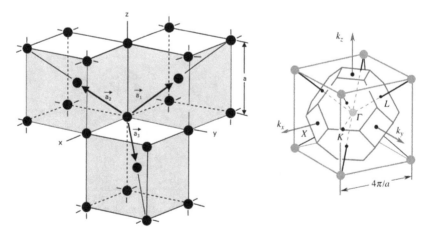

Figure 1.18. LEFT: BCC primitive translation vectors. All three vectors point to body centres in three different conventional cells (reproduced with permission from [2] John Wiley & Sons). RIGHT: the Wigner–Seitz cell for BCC is a truncated octahedron. The eight hexagonal faces bisect the lines between the body centre and the vertices. The six square faces bisect the lines between the body centre and the nearby body centres (reproduced from [3], copyright (2010) with permission from Springer Nature).

- Note that each point at each corner is shared by eight cells, so the corner point 'contributes' $\frac{1}{8}$ of a point to each cell and there is one point at the centre. Thus, in total, BCC has two points per cell.
- There are many examples of elements that occur in the BCC structure.

Face-centred cubic (FCC): The lattice has one point at each corner of the cube and one point at the (centre of the) face of the cube.
- The conventional cell is not the primitive cell. The conventional cell has four times the volume of the primitive cell.
- Taking one of the corner points as origin, the primitive translation vectors are $\vec{a}_1 = \frac{a}{2}(\hat{y} + \hat{z})$, $\vec{a}_2 = \frac{a}{2}(\hat{x} + \hat{z})$, and $\vec{a}_3 = \frac{a}{2}(\hat{x} + \hat{y})$. See figure 1.19.
- Note that each point at the corner is shared by eight cells, so each corner point 'contributes' $\frac{1}{8}$ of a point to each cell and there is one point (shared by two cells) at each of the six faces. Thus, in total, FCC has four points per cell.
- There are also many examples of elements that occur in the FCC structure.

A summary of the properties of cubic structures is shown in table 1.1.

1.1.3.4 Simple 3D lattices: the hexagonal lattice
Simple hexagonal: This lattice is shown in figure 1.20 but it doesn't actually exist in nature. This lattice is defined so that we can discuss the hexagonal close-packed structure in the next section.

1.1.3.5 The packing of hard spheres
For a solid to be stable, its atoms should be packed as close together as possible so that their potential energy is reduced. The atoms are also packed according to

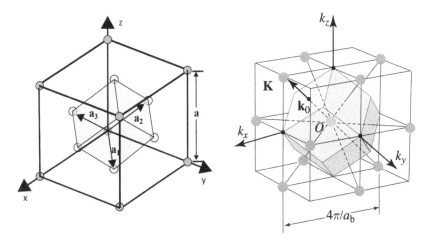

Figure 1.19. LEFT: FCC primitive translation vectors. All three vectors point to face centres in the same conventional cell (reproduced with permission from [2] John Wiley & Sons). RIGHT: the Wigner–Seitz cell for FCC is a rhombic dodecahedron. Note that the cube is not the conventional FCC cell. Each of the 12 (congruent) faces bisects a line that joins a corner point to a face-centred point (reproduced from [3], copyright (2010 with permission from Springer Nature).

Table 1.1. This table summarises the main properties of the three cubic structures. The packing fraction (or efficiency) is the maximum fraction of the available volume that can be filled with hard spheres. More detail on packing hard spheres will be given later.

Property	SC	BCC	FCC
Volume of the conventional cell	a^3	a^3	a^3
Lattice points per cell	1	2	4
Volume of the primitive cell	a^3	$\frac{1}{2}a^3$	$\frac{1}{4}a^3$
Lattice points per volume	$\frac{1}{a^3}$	$\frac{2}{a^3}$	$\frac{4}{a^3}$
No. of nearest neighbours (nn)	6	8	12
Nearest neighbour distance	a	$\frac{a\sqrt{3}}{2} = 0.866a$	$\frac{a}{\sqrt{2}} = 0.707a$
No. of next-nearest neighbours (nnn)	12	6	6
Next-nearest neighbour distance	$a\sqrt{2}$	a	a
Packing fraction (or efficiency)	$\frac{\pi}{6}$	$\frac{\pi\sqrt{3}}{8}$	$\frac{\pi\sqrt{2}}{6}$

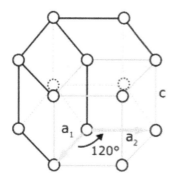

Figure 1.20. The simple hexagonal lattice has three primitive vectors: $\vec{a}_1 = -\frac{a}{2}\hat{x} - \frac{a\sqrt{3}}{2}\hat{y}$, $\vec{a}_2 = a\hat{y}$, and $\vec{a}_3 = c\hat{z}$.

Bravais lattice arrangements. Thus, we now want to see which types of Bravais lattice arrangements allow higher packing efficiencies and make the solid more stable.

We approximate atoms as hard (incompressible) spheres. We will then calculate the packing efficiencies for each Bravais lattice arrangement. We will discuss three types of cubic packing and a fourth type called hexagonal close packing (HCP).

Simple cubic packing (SCP): this type of packing simply means that the second layer of spheres is directly on top of the first layer. Let us calculate its packing efficiency (figure 1.21):

- Let the radius of a sphere be R and the side of the cube be a. So $a = 2R$.
- The packing efficiency is defined as $\dfrac{\text{volume of cube occupied by spheres}}{\text{volume of cube}} = \dfrac{1 \times \frac{4}{3}\pi R^3}{a^3}$
  $= \frac{\pi}{6} = 52\%$.

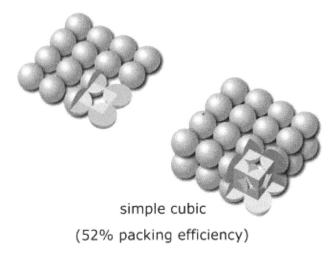

simple cubic

(52% packing efficiency)

Figure 1.21. Simple cubic packing (SCP).

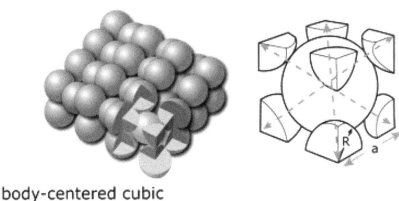

body-centered cubic
(68% packing efficiency)

Figure 1.22. Body-centred cubic packing (BCCP).

Body-centred cubic packing (BCCP): this type of packing simply means that the second layer of spheres sits in the holes of the first layer, and the third layer is directly on top of the first layer. Let us calculate its packing efficiency:

- Let the radius of a sphere be R and the side of the cube be a. The relationship between a and R is now not as simple. See the right figure above; note that the diagonal of the cube is $4R$ and is also $a\sqrt{3}$, so $a = \frac{4R}{\sqrt{3}}$ (ffigure 1.22).

- The packing efficiency is defined as $\frac{\text{volume of cube occupied by spheres}}{\text{volume of cube}} = \frac{2 \times \frac{4}{3}\pi R^3}{a^3}$ $= \frac{\pi\sqrt{3}}{8} = 68\%$.

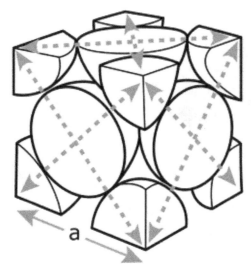

Figure 1.23. Face-centred cubic packing (FCCP).

Face-centred cubic packing (FCCP): this type of packing is not performed by stacking regular layers of spheres, so it is harder to describe and visualise. We will discuss it further when we compare it to HCP later. Let us calculate its packing efficiency:

- Let the radius of a sphere be R and the side of the cube be a. The relationship between a and R is now not that simple. See figure 1.23; note that the face diagonal of the cube is $4R$ and is also $a\sqrt{2}$, so $a = \frac{4R}{\sqrt{2}}$.

- The packing efficiency is defined as $\frac{\text{volume of cube occupied by spheres}}{\text{volume of cube}} = \frac{4 \times \frac{4}{3}\pi R^3}{a^3}$ $= \frac{\pi\sqrt{2}}{6} = 74\%$.

- We can see that this packing efficiency is the highest. This FCCP is also called cubic close packing (CCP), as it is the maximal packing configuration.

Hexagonal close packing (HCP): This type of packing is simply another way to achieve the maximal packing configuration in which the third layer is the same as the first layer.

- Note that HCP can be viewed as the simple hexagonal lattice with a two-atom basis.

- The two basis atoms are at $0\vec{a}_1 + 0\vec{a}_2 + 0\vec{a}_3$ and $\frac{1}{3}\vec{a}_1 + \frac{2}{3}\vec{a}_2 + \frac{1}{2}\vec{a}_3$ with respect to the simple hexagonal lattice points.

- The packing efficiency is the same as that of CCP, which is 74% or, more accurately, $\frac{\pi\sqrt{2}}{6}$. A question in the exercises asks the reader to show that the 74% packing efficiency is achieved when $\frac{c}{a} = \sqrt{\frac{8}{3}}$ (figures 1.24–1.27).

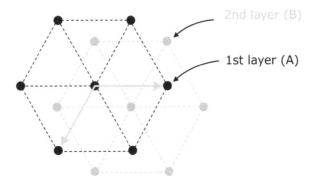

Figure 1.24. A top view of hexagonal close packing (HCP).

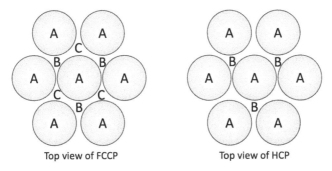

Top view of FCCP   Top view of HCP

Figure 1.25. A comparison of the top views of FCCP (or CCP) and HCP. The difference between them lies in the layering pattern: FCCP is an ABC type of layering, and HCP is an ABA type of layering. Note that this top view (left) is actually the (111) direction of the FCC cube in figure 1.23.

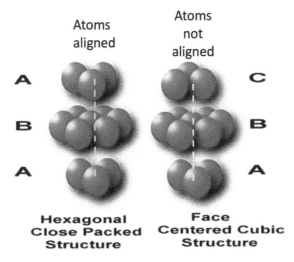

Figure 1.26. A comparison between the side views of FCCP (or CCP) and HCP. Reprinted with permission from Rodolfo Novakovic [4], Copyright (2006).

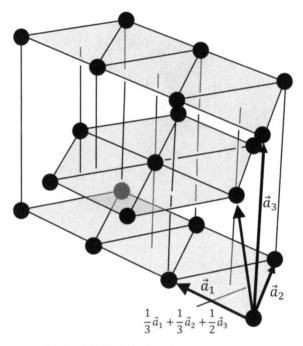

$$\frac{1}{3}\vec{a}_1 + \frac{1}{3}\vec{a}_2 + \frac{1}{2}\vec{a}_3$$

Figure 1.27. Another perspective, in which HCP is viewed as two interpenetrating simple hexagonal Bravais lattices displaced vertically by $\frac{c}{2}$ and displaced horizontally so that the points of the upper simple hexagonal lattice lie directly above the centres of the triangles of the lower simple hexagonal lattice [5]. Copyright (2006) John Wiley & Sons.

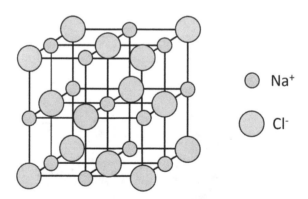

Figure 1.28. The structure of sodium chloride; where each side has length a.

1.1.4 A standard library of 3D structures

1.1.4.1 The sodium chloride structure (figure 1.28)

- The chemical formula is NaCl, so Na and Cl are in a 1:1 ratio in the crystal.

- Its structure is FCC and its basis is two dissimilar atoms at $(0, 0, 0)$ and $\left(\frac{1}{2}, \frac{1}{2}, \frac{1}{2}\right)$.[4]
- A second perspective is to think of NaCl as two interpenetrating FCC lattices. One FCC lattice is made of sodium ions only and the other is made of chloride ions only.

1.1.4.2 The caesium chloride structure (figure 1.29)
- Its structure is SC and its basis is two dissimilar atoms at $(0, 0, 0)$ and $\left(\frac{1}{2}, \frac{1}{2}, \frac{1}{2}\right)$.[5]

1.1.4.3 The calcium fluoride structure (figure 1.30)
- Its structure is SC and its basis is four calcium ions at $(0, 0, 0)$, $\left(0, \frac{1}{2}, \frac{1}{2}\right)$, $\left(\frac{1}{2}, 0, \frac{1}{2}\right)$, and $\left(\frac{1}{2}, \frac{1}{2}, 0\right)$ and eight fluoride ions at $\left(\frac{1}{4}, \frac{1}{4}, \frac{1}{4}\right)$, $\left(\frac{1}{4}, \frac{3}{4}, \frac{3}{4}\right)$, $\left(\frac{3}{4}, \frac{1}{4}, \frac{3}{4}\right)$, $\left(\frac{3}{4}, \frac{3}{4}, \frac{1}{4}\right)$, $\left(\frac{3}{4}, \frac{3}{4}, \frac{3}{4}\right)$, $\left(\frac{3}{4}, \frac{1}{4}, \frac{1}{4}\right)$, $\left(\frac{1}{4}, \frac{3}{4}, \frac{1}{4}\right)$, and $\left(\frac{1}{4}, \frac{1}{4}, \frac{3}{4}\right)$.

- A second perspective is to see its structure as FCC with a three-atom basis. The three-atom basis is: Ca at $(0, 0, 0)$ and F at $\left(\frac{1}{4}, \frac{1}{4}, \frac{1}{4}\right)$ and $\left(\frac{3}{4}, \frac{3}{4}, \frac{3}{4}\right)$.

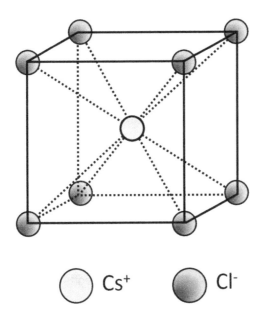

$\bigcirc$ Cs$^+$ $\bullet$ Cl$^-$

Figure 1.29. The caesium chloride structure; each side has length a.

---

[4] The notation $\left(\frac{1}{2}, \frac{1}{2}, \frac{1}{2}\right)$ is short for $\frac{1}{2}a\hat{x} + \frac{1}{2}a\hat{y} + \frac{1}{2}a\hat{z}$.
[5] You should not consider this to be BCC because the atoms at the corners and the atom at the body centre are not the same.

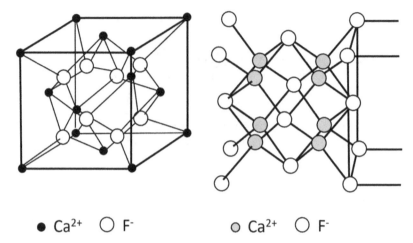

• Ca²⁺ ◯ F⁻   ◎ Ca²⁺ ◯ F⁻

Figure 1.30. The calcium fluoride structure; each side has length a. The picture on the right clearly shows (in colour) the five calcium atoms on a face.

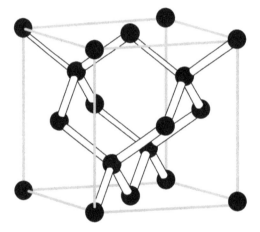

Figure 1.31. Diamond structure where each side has length a. All the atoms are identical.

1.1.4.4 The diamond structure (figure 1.31)

- Its structure is FCC and its basis is two identical atoms at $(0, 0, 0)$ and $\left(\frac{1}{4}, \frac{1}{4}, \frac{1}{4}\right)$.

- A second perspective is to see it as two interpenetrating FCC lattices (of identical atoms) displaced by $\frac{1}{4}$ along the body diagonal.

1.1.4.5 The zincblende (zinc sulphide) structure (figure 1.32)

- This structure is similar to that of diamond. Here, the structure is FCC with a basis of two atoms, which are simply two different atoms.
- A technologically very important compound that has this structure is GaAs.

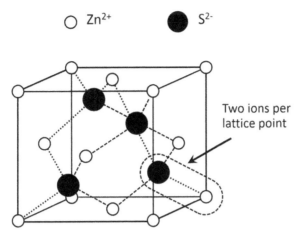

Figure 1.32. The zincblende structure; each side has length *a*. This structure is similar to that of diamond.

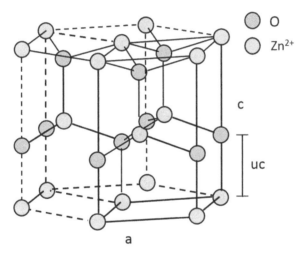

Figure 1.33. The wurtzite structure is a simple hexagonal structure with a four-atom basis. Note that length *c* is the entire vertical length.

- Note that the diamond structure has inversion symmetry about $\left(\frac{1}{8}, \frac{1}{8}, \frac{1}{8}\right)$ that the zincblende structure does not have, due to its basis of two different atoms.

1.1.4.6 The wurtzite (zinc oxide) structure (figure 1.33)
- Wutzite has a simple hexagonal structure with a basis of four atoms: two ions of zinc at (0, 0, 0) and $\left(\frac{a}{2}, \frac{a\sqrt{3}}{2}, \frac{c}{2}\right)$ and two oxygen ions at (0, 0, *uc*) and $\left(\frac{a}{2}, \frac{a\sqrt{3}}{2}, \frac{c}{2} + uc\right)$.

- A second perspective is to see it as two interpenetrating HCP lattices separated by (0, 0, *uc*). Usually, each HCP lattice is made up of one type of atom only.

1.1.4.7 The perovskite (calcium titanate) structure (figure 1.34)

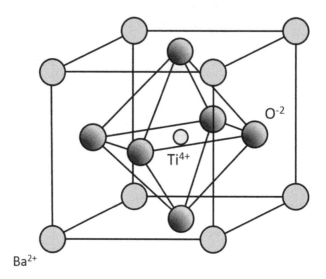

Figure 1.34. The perovskite structure is made out of three types of atoms: one type occupies the corners, one type occupies the faces, and the last type occupies the body centre.

1.2 Wave diffraction and the reciprocal lattice

1.2.1 X-ray diffraction of crystals

Since we cannot really look at solids and crystals with the microscope (or at least, it was not possible prior to the invention of the atomic force microscope in 1981), we have to determine the structures and properties of solids and crystals indirectly using scattering methods. The spacing between atoms in a crystal is a few angstroms, so we need waves that have wavelengths of that order to probe the crystal. Thus, x-rays are the most suitable waves for this purpose.

We could also use matter waves, such as electrons (of roughly 150 eV) or neutrons (of roughly 0.08 eV), to probe the crystal[6].

When x-ray diffraction experiments are conducted (see figure 1.35), the structure of the crystal is deduced from the intensities of the x-rays and their corresponding diffraction angles.

1.2.1.1 Bragg's law

A simple analysis of x-ray diffraction makes these two assumptions:
   1. Diffraction peaks are caused by the constructive interference of waves (of x-rays) reflected from parallel planes of atoms.

---

[6] These are wave scattering processes. There are two types of scattering: elastic (in which energy is conserved and there is only a change of direction in momentum) and inelastic (in which energy is not conserved).

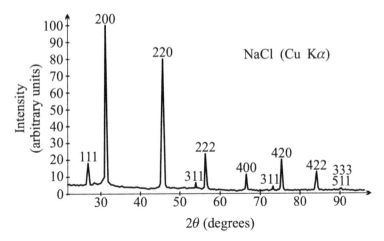

Figure 1.35. X-ray diffraction data for sodium chloride in terms of intensity vs diffraction angle. Reprinted by permission from Springer Nature [6], Copyright (2007).

2. The reflection is specular, which means we can further assume that the angle of incidence is equal to the angle of reflection.

We can use these two assumptions to derive William Bragg's law of x-ray diffraction,[7] which describes x-ray diffraction due to a Bravais lattice:

$$\text{constructive interference means: path difference}$$
$$= \text{integer multiples of wavelength} \tag{1.5}$$

$$2d \sin \theta = n\lambda \text{ where } n \text{ is an integer} \tag{1.6}$$

This is Bragg's law (figure 1.36), for which we can make the following comments:
- The perfect periodicity of the Bravais lattice is important so that we can talk about planes in the solid.
- The concept of planes previously led us to the concept of the reciprocal lattice. So we shall use the concept of reciprocal space to describe x-ray diffraction extensively later.
- Note that this law for the maximum intensity of x-rays did not take into account the basis 'decorating' each point. It turns out that the basis also affects the intensity, as we will discuss later.

1.2.1.2 Laue's condition
Max von Laue (1912) offered a more elegant discussion of x-ray diffraction, which we will turn to now. It turns out that Laue's condition and Bragg's law are equivalent.

---

[7] William Henry Bragg and his son William Lawrence Bragg were jointly awarded the Nobel Prize in Physics for their contribution to x-ray diffraction. This is the only instance of father–son awardees in the history of Nobel Prizes.

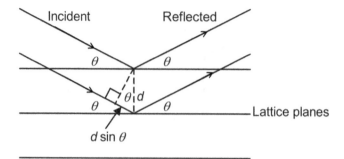

Figure 1.36. X-ray diffraction in a crystal. The distance d is the distance between two parallel planes. This distance is usually not the spacing of the lattice points but the distance between planes of atoms. The incident angle is θ. Reprinted by permission from Springer Nature [7], Copyright (2018).

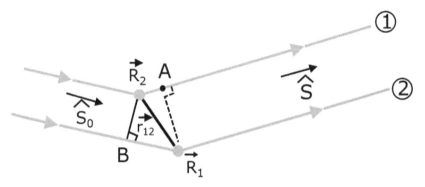

Figure 1.37. The unit vector $\hat{s}_0$ is in the direction of the incident wave and the unit vector $\hat{s}$ is in the direction of the scattered wave. Let $\vec{R}_1$ and $\vec{R}_2$ be the position vectors of a pair of atoms so that $\vec{r}_{12} = \vec{R}_1 - \vec{R}_2$ is the relative vector. Reprinted by permission from Springer Nature [8], Copyright (2018).

Max von Laue considered the scattering of x-rays by a pair of lattice points in a crystal as shown in figure 1.37:

For constructive interference, the path difference = integer multiples of λ.

$$m\lambda = \text{path difference} \tag{1.7}$$

$$m\lambda = |R_2 A - BR_1| \tag{1.8}$$

$$m\lambda = |\vec{r}_{12} \cdot \hat{s} - \vec{r}_{12} \cdot \hat{s}_0| \tag{1.9}$$
| assuming the dot product is positive

$$m\lambda = \vec{r}_{12} \cdot (\hat{s} - \hat{s}_0)$$
| note that $\vec{r}_{12}$ must be a translation vector $\vec{R} = n_1 \vec{a}_1 + n_2 \vec{a}_2 + n_3 \vec{a}_3$ then write $\tag{1.10}$

$ph\lambda, pk\lambda, pl\lambda = \vec{a}_1 \cdot (\hat{s} - \hat{s}_0), \; \vec{a}_2 \cdot (\hat{s} - \hat{s}_0), \; \vec{a}_3 \cdot (\hat{s} - \hat{s}_0)$

| where $h = \dfrac{m}{3pn_1}, \; k = \dfrac{m}{3pn_2}, \; l = \dfrac{m}{3pn_3}$ are integers　　(1.11)

| and where p is a common multiplier

| we can write as components in reciprocal bases $\vec{b}_1, \vec{b}_2 \; and \; \vec{b}_3$

$(\hat{s} - \hat{s}_0) = (\vec{a}_1 \cdot (\hat{s} - \hat{s}_0))\vec{b}_1 + (\vec{a}_2 \cdot (\hat{s} - \hat{s}_0))\vec{b}_2 + (\vec{a}_3 \cdot (\hat{s} - \hat{s}_0))\vec{b}_3$

(1.12)

| this works if we recall $\vec{a}_i \cdot \vec{b}_j = 2\pi\delta_{ij}$ and test by taking the dot product with $\vec{a}_i$

$2\pi(\hat{s} - \hat{s}_0) = p\lambda(h\vec{b}_1 + k\vec{b}_2 + l\vec{b}_3)$

(1.13)

| recall $\vec{G} = h\vec{b}_1 + k\vec{b}_2 + l\vec{b}_3$ is the reciprocal space translation vector

$\dfrac{2\pi(\hat{s} - \hat{s}_0)}{\lambda} = p\vec{G}$

| note that $\dfrac{2\pi}{\lambda} = k$, which is the magnitude of the wavevector　　(1.14)

| thus write $\vec{k} = \dfrac{2\pi}{\lambda}\hat{s}$ and $\vec{k}_0 = \dfrac{2\pi}{\lambda}\hat{s}_0$ and $\Delta\vec{k} = \vec{k} - \vec{k}_0$

$\Delta\vec{k} = p\vec{G}$

(1.15)

| note that 'integer $p \times \vec{G}$' is just another translation vector $\vec{G}$

$$\boxed{\Delta\vec{k} = \vec{G}}$$
(1.16)

This is the Laue equation or Laue's condition for constructive interference. It means that constructive interference occurs if the change in wavevector is a reciprocal translation vector $\vec{G}$ (figure 1.37).

We shall proceed with a third way of writing diffraction conditions. The idea is to recall that the vector $\vec{G}$ is normal to the (hkl) plane (in real space). See figure 1.38 for the derivation of the third way of writing diffraction conditions:

$$\boxed{2\vec{k} \cdot \vec{G} = \left|\vec{G}\right|^2.}$$
(1.17)

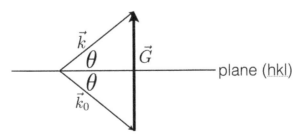

Figure 1.38. Recall that $\vec{k}_0$ and $\vec{k}$ have the same magnitude. So $\vec{k} - \frac{1}{2}\vec{G}$ is a vector in the plane. Thus $\left(\vec{k} - \frac{1}{2}\vec{G}\right) \cdot \vec{G} = 0$, which leads to $2\vec{k} \cdot \vec{G} = |\vec{G}|^2$, the third way of writing the diffraction condition.

Proving that Laue's condition and Bragg's law are equivalent:

If you examine the above derivation of Laue's condition carefully, you will realise that it contains the same physical assumptions as Bragg's law; thus, we suspect that they are actually the same.

Here we shall show that easily using figure 1.38. Recall that $d = \frac{2\pi}{|\vec{G}|}$ and from figure 1.38, $|\vec{G}| = 2k \sin\theta$; then, writing $k = \frac{2\pi}{\lambda}$ leads to $2d \sin\theta = \lambda$, which is indeed Bragg's law.[8]

1.2.1.3 Ewald's construction

This is a geometrical construction (in reciprocal space) that allows x-ray diffraction conditions to be displayed and fulfilled by simple geometrical drawings. Figure 1.38 gives some clues to this construction:

- The incident and scattered wavevectors have the same magnitude and thus the same length. Geometrically, the two wavevectors are radii of a sphere. Their magnitude is $\frac{2\pi}{\lambda}$.
- The vector difference between the two wavevectors must be a reciprocal space translation vector $\vec{G}$; this means that the two arrowheads of the two wavevectors must touch a point in reciprocal space.

These conditions are not easy to fulfil all at the same time, especially in 3D!

Ewald's construction proceeds with the following steps:

1. Choose the origin such that $\vec{k}_0$ terminates at any reciprocal lattice point.
2. Draw a sphere (a circle in 2D) of radius $|\vec{k}_0| = \frac{2\pi}{\lambda}$ about the origin. This is called the Ewald sphere.
3. Look for other reciprocal lattice points that lie on the surface of the Ewald sphere. For each such lattice point, draw $\vec{k}$.
4. The angle between $\vec{k}_0$ and $\vec{k}$ is 2θ, and the vector $\vec{G} = \vec{k} - \vec{k}_0$ tells us the (hkl) plane involved in the diffraction (figure 1.39).

1.2.1.4 The geometrical structure factor and the atomic form factor

We now discuss how the basis 'decorating' each lattice point affects the x-ray diffraction intensity. The types of atoms in the basis also affect the x-ray diffraction intensity. Therefore, we have two factors to consider.

---

[8] The 'missing' integer m occurs simply because h, k, l in $\vec{G}$ were assumed to be in the simplest ratio. Similarly, $\vec{G}$ is just integer multiples of the simplest ratio $\vec{G}$.

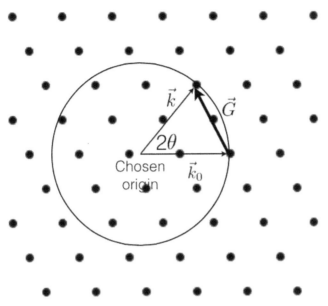

Figure 1.39. The construction of a Ewald sphere to search for diffraction peaks. Note that the origin does not coincide with a reciprocal lattice point.

Geometrical structure factor: the geometrical structure factor considers how the arrangement of the basis affects the intensity of the x-ray diffraction pattern.

- Think of the Bravais lattice as having n (identical) atoms as the basis. Their displacement vectors (from the Bravais lattice point) are $\vec{d}_1, \vec{d}_2, \ldots, \vec{d}_n$.
- Looking back at Laue condition, we borrow the expression for the 'path difference':

$$\text{path difference} = \vec{r}_{12} \cdot (\hat{s} - \hat{s}_0) \tag{1.18}$$

$$| \text{ in the context of basis atoms, write } \vec{r}_{12} = \vec{d}_1 - \vec{d}_2$$

$$\text{path difference} = (\vec{d}_1 - \vec{d}_2) \cdot (\hat{s} - \hat{s}_0) \tag{1.19}$$

$$| \text{ then write } \vec{k} = \frac{2\pi}{\lambda}\hat{s} \text{ and } \vec{k}_0 = \frac{2\pi}{\lambda}\hat{s}_0 \text{ and } \Delta\vec{k} = \vec{k} - \vec{k}_0$$

$$\text{path difference} = \frac{\lambda}{2\pi}(\vec{d}_1 - \vec{d}_2) \cdot \Delta\vec{k} \tag{1.20}$$

$$| \text{ recall that 'phase difference'} = \text{'path difference'} \times \frac{2\pi}{\lambda}$$

$$\text{phase difference} = (\vec{d}_1 - \vec{d}_2) \cdot \Delta\vec{k} \tag{1.21}$$

$$| \text{ recall that the Laue condition says } \Delta\vec{k} = \vec{G}$$

$$\text{phase difference} = (\vec{d}_1 - \vec{d}_2) \cdot \vec{G} \tag{1.22}$$

- Considering the complex form of waves: $Ae^{i\vec{k}\cdot\vec{r}-i\omega t}$, the phase difference means that the two waves have a ratio of $e^{i(\vec{d}_1-\vec{d}_2)\cdot\vec{G}}$.
- Extrapolating, the ratio between waves scattered from basis atoms two and three is $e^{i(\vec{d}_2-\vec{d}_3)\cdot\vec{G}}$.
- So we can write the relative phases scattered from each basis atom $\vec{d}_1, \ldots, \vec{d}_n$ as $Ae^{i\vec{G}\cdot\vec{d}_1}, \ldots, Ae^{i\vec{G}\cdot\vec{d}_n}$, where A is some constant.
- The total amplitude is just the sum of waves which have relative phases $e^{i\vec{G}\cdot\vec{d}_1}, \ldots, e^{i\vec{G}\cdot\vec{d}_n}$. This is called the geometrical structure factor $S_{\vec{G}}$

$$S_{\vec{G}} = A\sum_{i=1}^{n} e^{i\vec{G}\cdot\vec{d}_i} \tag{1.23}$$

- Let us test this factor on three important examples:
  1. SC: this example might be a joke, since the simple cubic is a one-atom basis and the single atom is at the Bravais lattice point. So $n = 1$, and $\vec{d}_1 = 0\hat{x} + 0\hat{y} + 0\hat{z}$; thus $S_{\vec{G}} = Ae^0 = A$.
  2. BCC: we treat BCC as an SC Bravais lattice with a two-atom basis, so $n = 2$ and displacement vectors $\vec{d}_1 = 0\hat{x} + 0\hat{y} + 0\hat{z}$ and $\vec{d}_2 = \frac{a}{2}(\hat{x} + \hat{y} + \hat{z})$.
     - The reciprocal space of SC is also SC; the general reciprocal translation vector (see figure 1.42) is $\vec{G} = h\vec{b}_1 + k\vec{b}_2 + l\vec{b}_3 = \frac{2\pi}{a}(h\hat{x} + k\hat{y} + l\hat{z})$.
     - Finally, the geometrical structure factor is

$$S_{\vec{G}} = A\left(e^{i\frac{2\pi}{a}(0h+0k+0l)} + e^{i\frac{2\pi}{a}(h\frac{a}{2}+k\frac{a}{2}+l\frac{a}{2})}\right) \tag{1.24}$$

$$=A(1 + e^{i\pi(h+k+l)}) \tag{1.25}$$

$$=A(1 + (e^{i\pi})^{h+k+l}) \tag{1.26}$$

$$=A(1 + (-1)^{h+k+l}) \tag{1.27}$$

$$S_{\vec{G}} = \begin{cases} 2A \text{ if } h + k + l = \text{even} \\ 0 \text{ if } h + k + l = \text{odd} \end{cases} \tag{1.28}$$

It is interesting to examine this geometrical structure factor more closely. The geometrical structure factor can actually be zero for certain (hkl) planes, and we should just remove these hkl points from the SC reciprocal space. This will be done later as an exercise for you to find the resulting structure of the reciprocal space when we remove points satisfying $h + k + l = $ odd from the SC reciprocal space.

3. FCC: this case is used in an exercise later; I will just state the results here. Assuming FCC is SC with a four-atom basis whose displacement vectors are: $\vec{d}_1 = 0\hat{x} + 0\hat{y} + 0\hat{z}$, $\vec{d}_2 = \frac{a}{2}\hat{x} + \frac{a}{2}\hat{y} + 0\hat{z}$, $\vec{d}_3 = \frac{a}{2}\hat{x} + 0\hat{y} + \frac{a}{2}\hat{z}$, and $\vec{d}_4 = 0\hat{x} + \frac{a}{2}\hat{y} + \frac{a}{2}\hat{z}$, the geometrical structure factor is

$$S_{\vec{G}} = \begin{cases} 4A \text{ if } h, k, l \text{ are all even or all odd} \\ 0 \text{ if } h, k, l \text{ are partially even and partially odd} \end{cases} \qquad (1.29)$$

Atomic form factor: The atomic form factor considers how, if the atoms in the basis are not identical, the electron densities of the different atoms affect the x-ray diffraction intensity. We modify the amplitude factor and make it dependent on the basis atom (and change the notation from A to f_i).

$$S_{\vec{G}} = \sum_{i=1}^{n} f_i(\vec{G}) e^{i\vec{G}\cdot\vec{d}_i} \qquad (1.30)$$

$f_i(\vec{G})$ is called the atomic form factor. Of course, if all the atoms are the same, then all $f_i(\vec{G})$ factors are the same. We shall not discuss the atomic form factor further except to say that it is now highly unlikely that $S_{\vec{G}}$ can be zero.

1.2.2 Reciprocal space (the full story)

Earlier, we introduced the reciprocal space as a 'nice' way to keep track of planes in real space. Let us just summarise it as 'Perspective 1':

Perspective 1 of reciprocal space:

For x-ray diffraction, we really care about which planes are involved in making those bright x-ray patterns on the screen. To focus on the planes, we shall invent a space called the reciprocal space, in which every point represents a set of parallel planes in the real crystal lattice.

The (3D) primitive translation vectors of reciprocal space are defined as follows:

$$\vec{b}_1 = 2\pi\frac{\vec{a}_2 \times \vec{a}_3}{|\vec{a}_1 \cdot \vec{a}_2 \times \vec{a}_3|}, \quad \vec{b}_2 = 2\pi\frac{\vec{a}_3 \times \vec{a}_1}{|\vec{a}_1 \cdot \vec{a}_2 \times \vec{a}_3|}, \quad \vec{b}_3 = 2\pi\frac{\vec{a}_1 \times \vec{a}_2}{|\vec{a}_1 \cdot \vec{a}_2 \times \vec{a}_3|} \qquad (1.31)$$

with $\vec{a}_i \cdot \vec{b}_j = 2\pi\delta_{ij}$. A general translation vector in reciprocal space is

$$\vec{G} = h\vec{b}_1 + k\vec{b}_2 + l\vec{b}_3, \quad \text{where } h, k, l \text{ are integers} \qquad (1.32)$$

It turns out that the concept of reciprocal space is deeper than that. This is 'Perspective 2', and we will expand on it here. It also turns out that the concept of reciprocal space is much more useful than keeping track of planes. It will also be used in chapter 3 to describe electron waves in a solid.

Perspective 2 of reciprocal space:

Due to the periodicity of the real crystal lattice, a function in real space can be written as a Fourier series in terms of the reciprocal space translation vector $\vec{G}$.

The opposite is true as well: due to the periodicity of the reciprocal space, a function in reciprocal space can be written as a Fourier series in terms of the real space translation vector $\vec{R}$.

These two Fourier series are the inverses of each other.

We now digress slightly to revise and write out the Fourier series explicitly. For a **1D Fourier series,** consider a 1D system in which a function $f(x)$ is periodic and has the period a; see figure 1.40, for example.

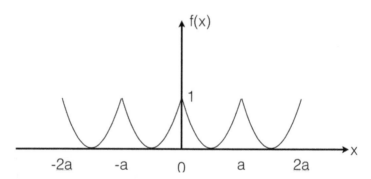

Figure 1.40. A 1D periodic function with period a.

The 1D real space translation vector is $\vec{R} = n_1\vec{a}_1$ with $\vec{a}_1 = a\hat{x}$. The reciprocal space translation vector is $\vec{G} = h\vec{b}_1$ with $\vec{b}_1 = \frac{2\pi}{a}\hat{x}$.

This periodicity means that $f(x) = f(x + n_1 a)$, which means the function looks the same when translated by $n_1 a$.

We shall check that the correct Fourier expansion of $f(x)$ (in exponential form) is

$$f(x) = \sum_{h=-\infty}^{+\infty} f_h \, e^{i\frac{2\pi h}{a}x}$$

(1.33)

| where f_h are (complex) Fourier coefficients

We check by starting with: $f(x + n_1a) = \sum\limits_{h=-\infty}^{+\infty} f_h\, e^{i\frac{2\pi h}{a}(x+n_1a)}$ (1.34)

$$f(x + n_1a) = \sum\limits_{h=-\infty}^{+\infty} f_h\, e^{i\frac{2\pi h}{a}x} e^{i\frac{2\pi h}{a}n_1a}$$ (1.35)

| note that $e^{i\frac{2\pi h}{a}n_1a} = e^{i2\pi hn_1} = 1$

$$f(x + n_1a) = \sum\limits_{h=-\infty}^{+\infty} f_h\, e^{i\frac{2\pi h}{a}x}$$ (1.36)

$f(x + n_1a) = f(x)$ indeed

Very importantly, note: $f(x) = \sum\limits_{h=-\infty}^{+\infty} f_h\, e^{i\frac{2\pi h}{a}x} = \sum\limits_{h=-\infty}^{+\infty} f_h\, e^{i\vec{G}\cdot\vec{x}}$ (1.37)

This last expression $f(x) = \sum_{h=-\infty}^{+\infty} f_h\, e^{i\vec{G}\cdot\vec{x}}$ shows the Fourier relationship between the two spaces, as stated earlier. The relationship is even clearer when we check the other way round: assume a function $g(k_x)$ that is periodic in reciprocal space with period $\frac{2\pi}{a}$, so that $g\left(k_x + h\frac{2\pi}{a}\right) = g(k_x)$. The correct Fourier expansion is then $g(k_x) = \sum_{n_1=-\infty}^{+\infty} g_{n_1}\, e^{i\vec{R}\cdot\vec{k}_x}$.

3D Fourier series: we will not redo the derivations but just state the expressions in 3D.

Real space translation vector: $\vec{R} = n_1\vec{a}_1 + n_2\vec{a}_2 + n_3\vec{a}_3$

Reciprocal space translation vector: $\vec{G} = h\vec{b}_1 + k\vec{b}_2 + l\vec{b}_3$. Note that $e^{i\vec{G}\cdot\vec{R}} = 1$

Real space periodic function: $f(\vec{r}) = f(\vec{r} + \vec{R})$ (1.38)

Fourier expansion of real space periodic function: $f(\vec{r}) = \sum\limits_{\vec{G}} f_{\vec{G}}\, e^{i\vec{G}\cdot\vec{r}}$

1.2.2.1 Brillouin zones (BZs)

In chapter 3, you will see that the reciprocal space is very important for describing electron energies in the solid. Specifically, the Wigner–Seitz construction of the reciprocal space primitive cell will turn out to contain all the physics required. This choice of primitive cell in reciprocal space shall be called a Brillouin zone.

We shall take a simple 2D Bravais lattice to illustrate the drawing of BZs in reciprocal space (figure 1.41).

1.2.2.2 Examples of reciprocal spaces and Brillouin zones

Here, we will list the real and reciprocal spaces of the cubic structures for reference.

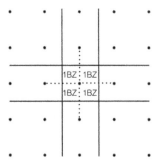 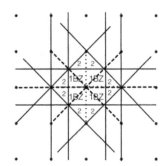

Reciprocal Space
Take the centre point as origin

Wigner-Seitz Construction:
1. Join from origin to the nearest neighbour (nn) points
2. Draw perpendicular bisector planes
3. Enclosed volume is 1st Brillouin Zone (1BZ)

Extend Construction to 2BZ:
1. Join from origin to the next nearest neighbour (nnn) points and further
2. Draw perpendicular bisector planes
3. Enclosed volume is 2nd Brillouin Zone (2BZ)

Figure 1.41. LEFT: the Wigner–Seitz construction in the reciprocal space gives the important first Brillouin Zone (1BZ). RIGHT: extension of the Wigner–Seitz construction to construct the second Brillouin Zone (2BZ). Note that the area of the 2BZ is the same as that of the 1BZ. In fact, all BZs have the same area.

Simple cubic example:

Recall the real−space primitive vectors: $\begin{cases} \vec{a}_1 = a\hat{x} \\ \vec{a}_2 = a\hat{y} \\ \vec{a}_3 = a\hat{z} \end{cases}$

So reciprocal space primitive vectors: $\begin{cases} \vec{b}_1 = \frac{2\pi}{a}\hat{x} \\ \vec{b}_2 = \frac{2\pi}{a}\hat{y} \\ \vec{b}_3 = \frac{2\pi}{a}\hat{z} \end{cases}$

Thus, the reciprocal space is a simple cubic structure, which is the same structure as the real space (figure 1.42).

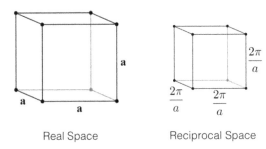

Real Space   Reciprocal Space

Figure 1.42. The real space simple cubic lattice and its reciprocal space, which is also simple cubic (not drawn to scale). Do you know how to draw 1BZ on the right figure?

Body-centred cubic example:

Recall the real–space primitive vectors: $\begin{cases} \vec{a}_1 = \frac{a}{2}(-\hat{x} + \hat{y} + \hat{z}) \\ \vec{a}_2 = \frac{a}{2}(\hat{x} - \hat{y} + \hat{z}) \\ \vec{a}_3 = \frac{a}{2}(\hat{x} + \hat{y} - \hat{z}) \end{cases}$

So reciprocal space primitive vectors: $\begin{cases} \vec{b}_1 = \frac{2\pi}{a}(\hat{y} + \hat{z}) \\ \vec{b}_2 = \frac{2\pi}{a}(\hat{x} + \hat{z}) \\ \vec{b}_3 = \frac{2\pi}{a}(\hat{x} + \hat{y}) \end{cases}$

Thus, the reciprocal space is a face-centred cubic structure! See figure 1.43.

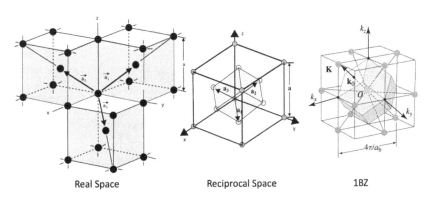

Real Space   Reciprocal Space   1BZ

Figure 1.43. The real space body-centred cubic lattice and its reciprocal space, which is face-centred in structure! The 1BZ is a rhombic dodecahedron. Reprinted by permission from [2] John Wiley & Sons and Springer Nature [3], Copyright (2010).

Face-centred cubic example:

Recall the real–space primitive vectors:
$$\begin{cases} \vec{a}_1 = \frac{a}{2}(\hat{y} + \hat{z}) \\ \vec{a}_2 = \frac{a}{2}(\hat{x} + \hat{z}) \\ \vec{a}_3 = \frac{a}{2}(\hat{x} + \hat{y}) \end{cases}$$

So reciprocal space primitive vectors:
$$\begin{cases} \vec{b}_1 = \frac{2\pi}{a}(-\hat{x} + \hat{y} + \hat{z}) \\ \vec{b}_2 = \frac{2\pi}{a}(\hat{x} - \hat{y} + \hat{z}) \\ \vec{b}_3 = \frac{2\pi}{a}(\hat{x} + \hat{y} - \hat{z}) \end{cases}$$

Thus the reciprocal space is a body-centred cubic structure! See figure 1.44.

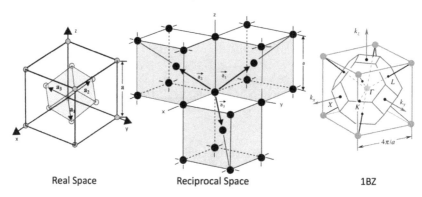

| Real Space | Reciprocal Space | 1BZ |

Figure 1.44. The real space face-centred cubic lattice and its reciprocal space, which is body-centred in structure! The 1BZ is a truncated octahedron. Reprinted by permission from [2] John Wiley & Sons and Springer Nature [3], Copyright (2010).

1.3 Crystal binding

In this section, we are concerned with the physics of how a crystal can be stable in the context of its internal forces. The concept of forces in a crystal is actually quite easy to understand. If there are only repulsive forces, then the crystal cannot possibly be stable. Obviously, the same is true when there are only attractive forces. Thus there must be both types of forces. In different types of solids, the physical origins of the attractive force and the repulsive force vary slightly.

In this section, we will just discuss very simple models of bonding and cohesion in solids. It will be useful to recall your existing knowledge of chemical bonding: van der Waals forces, ionic bonding, covalent bonding, and hydrogen bonding.

1.3.1 Inert/rare-gas solids

- These are made up of noble atoms, whose valence shells are filled (they belong to the last group of the periodic table). Their electron distribution is roughly spherically symmetric.
- The slight distortion of the electron distribution gives rise to van der Waals interaction, which is the attractive force.
- The repulsive force is due to the Pauli exclusion principle. When two atoms' electron distributions start to overlap, the principle forbids multiple occupancy of the same state by electrons. The electrons can occupy higher states, but energy is needed. Thus the principle discourages the overlap of electron distributions in the first place.
- Based on the above considerations, the interactions of a pair of rare-gas atoms can be modelled by the Lennard-Jones potential.

$$\phi(r) = 4\varepsilon\left[\left(\frac{\sigma}{r}\right)^{12} - \left(\frac{\sigma}{r}\right)^{6}\right] \tag{1.39}$$

A few comments are in order:

- The parameters ϵ and σ are constants to be fitted from experiments. From figure 1.45, we can say that ϵ characterises the strength of the attraction between the two atoms and σ characterises the radii of the repulsive cores.
- The variable r is the distance between the two atoms.
- The r^{-6} term represents attraction due to the van der Waals interaction. Recall from elementary physics that negative potential means attraction.

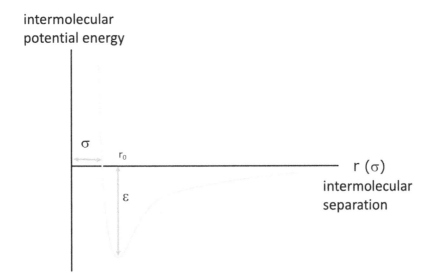

Figure 1.45. The Lennard-Jones potential is used to model the interatomic potential between a pair of rare-gas atoms. Note that the deeper the well depth ϵ, the stronger the intermolecular attraction.

- The r^{-12} term models repulsion due to the Pauli exclusion principle. Recall from elementary physics that positive potential means repulsion. This term only comes into effect when r is very small, as we recall from the earlier discussion that repulsion happens only when the atoms start to overlap.
- The total potential energy in the solid is the sum of all pairs of potential energy,

$$U(\vec{R}) = \frac{1}{2}N \sum_{\vec{R}(\neq 0)} \phi(\vec{R}) \qquad (1.40)$$

$$U(\vec{R}) = 2N\varepsilon \sum_{\vec{R}(\neq 0)} \left[\left(\frac{\sigma}{|\vec{R}|} \right)^{12} - \left(\frac{\sigma}{|\vec{R}|} \right)^{6} \right]$$

| the factor N appears because we are summing w.r.t. to a reference atom (1.41)

| the factor $\frac{1}{2}$ is presentto to undo the double−counting due to the summation

| write $\vec{R} = p(\vec{R})r$ where r is the nn distance, $p(\vec{R})$ is an integer

$$U(\vec{R}) = 2N\varepsilon \left[\left(\sum_{\vec{R}(\neq 0)} \frac{1}{p(\vec{R})^{12}} \right) \left(\frac{\sigma}{r} \right)^{12} - \left(\sum_{\vec{R}(\neq 0)} \frac{1}{p(\vec{R})^{6}} \right) \left(\frac{\sigma}{r} \right)^{6} \right] \qquad (1.42)$$

| we shall label $\displaystyle\sum_{\vec{R}(\neq 0)} \frac{1}{p(\vec{R})^{12}} = A_{12}$ and $\displaystyle\sum_{\vec{R}(\neq 0)} \frac{1}{p(\vec{R})^{6}} = A_{6}$

$$(\vec{R}) = 2N\varepsilon \left[A_{12} \left(\frac{\sigma}{r} \right)^{12} - A_{6} \left(\frac{\sigma}{r} \right)^{6} \right] \qquad (1.43)$$

A few comments are in order:
- The contributions from the nearest neighbours (NNs) are the largest. Note that for nn, $p(\vec{R}) = 1$.
- It is therefore sensible to make the approximation of summing over the NNs only.
- Let me give an example: if we have a FCC inert-gas solid, then summing over the NNs only, which for FCC are 12 in number, we get $A_{12} = \sum \frac{1}{1^{12}} = 12$ and $A_{6} = 12$.
- The equilibrium separation $r = r_0$ means that there is no net force or that the potential energy is at a minimum. This means we need to differentiate U and set its derivative to zero, i.e. $\frac{\partial U}{\partial r}\Big|_{r=r_0} = 0$. This calculation is left as an exercise; you should find that the equilibrium separation is $r_0 = \left(\frac{2A_{12}}{A_6} \right)^{1/6} \sigma$.
- Thus, the equilibrium energy (or cohesive energy) of the solid is (at zero temperature and zero pressure) $U = -\frac{N\varepsilon A_6^2}{2A_{12}}$ (exercise).
- One way to experimentally measure the values of ϵ and σ is to measure the bulk modulus, B. This is the resistance of the crystal to uniform compression.

It is defined as $B = -V\left(\frac{\partial P}{\partial V}\right)_T$. We will now carry out a thermodynamics analysis to write B in terms of U.

$$B = -V\left(\frac{\partial P}{\partial V}\right)_T$$

| the thermodynamic relation: $F = U - TS \Longrightarrow dF = dU - SdT - TdS$

| use the first law of thermodynamics: $dU = TdS - PdV$

| to get $dF = -PdV - SdT \Longrightarrow P = -\left(\frac{\partial F}{\partial V}\right)_T$

(1.44)

$$B = V\left(\frac{\partial^2 F}{\partial V^2}\right)_T$$

(1.45)

| take the simple case of $T = 0$ so $F = U$

$$B = V\left(\frac{\partial^2 U}{\partial V^2}\right)_{T=0}$$

(1.46)

Thus, the next step is to substitute the total potential energy U into equation (1.46); we can then determine the constants ϵ and σ from B.

1.3.2 Ionic crystals

- From basic chemistry, we know that for ionic bonding, electrons transfer between atoms and the atoms end up as charged cations and anions. Thus, the attractive force must be the (long-range) electrostatic force between ions that have charges of opposite signs. Obviously, the Coulomb potential is $\pm\frac{q_1 q_2}{4\pi\varepsilon_0 r}$, where '+' means repulsion and '−' means attraction. This electrostatic binding of ionic crystals will be called the Madelung energy.
- There are two contributions to the repulsive force: both Coulomb repulsion and the Pauli exclusion principle are in play between electrons when ions get close together. The net effect of repulsion will simply be modelled using an exponential form proposed by Born and Mayer, namely $\lambda e^{-r/\rho}$, where λ and ρ are experimental parameters. Note that this is not the only model for repulsion; other repulsion models are possible.
- The van der Waals interaction is weak and is thus negligible here.
- The total potential energy of the ionic crystal is the sum of the Madelung energy and the Born–Mayer energy. We shall make the usual NN approximation to the repulsive Born–Mayer energy, so the expression for the total energy is (assuming the two ions have the same magnitude q):

$$U = Nz\lambda e^{-R/\rho} - N\left(\sum_{i,j} \pm \frac{1}{P_{ij}}\right)\frac{q^2}{4\pi\varepsilon_0 R} = N\left[z\lambda e^{-R/\rho} - \alpha\frac{q^2}{4\pi\varepsilon_0 R}\right]$$

(1.47)

- where N is the number of cells,
- z is the number of NNs (in one cell),
- R is the NN distance,
- p_{ij} is an integer that denotes multiples of the NN distance R,
- and where $\alpha = \sum_{i,j} \pm \frac{1}{p_{ij}}$ is called the Madelung constant and we have to calculate it.
- There is no double counting here because we work with a particular cell i and sum over other j cells then multiply by N cells. Within cell i, we have to sum over z neighbours.
- Let us turn to a 1D example that shows the calculation of the Madelung constant α (figure 1.46). Here, one cell refers to one ion.

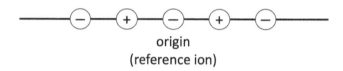

origin
(reference ion)

Figure 1.46. A one-dimensional ionic solid chain used to calculate the Madelung constant.

$$\alpha = \sum_{i,j} \pm \frac{1}{p_{ij}} \tag{1.48}$$

$$\alpha = 2 \times \left[\frac{1}{1} - \frac{1}{2} + \frac{1}{3} - \frac{1}{4} \cdots \right]$$

| the factor of two is due to two ions (left and right) at equal distances $\qquad$ (1.49)

| recall $\ln(1 + x) = x - \frac{1}{2}x^2 + \frac{1}{3}x^3 - \frac{1}{4}x^4 + \cdots$ so $x = 1$

$$\alpha = 2 \ln 2 \tag{1.50}$$

The calculations of the Madelung constant for 2D and 3D are complex, and we will not consider them in this book. Note that $\alpha > 0$ means that the crystal is stable (as U can be negative) and $\alpha < 0$ means that the crystal is unstable, as U is positive.

- The equilibrium separation $R = R_0$ is again calculated by $\frac{\partial U}{\partial R}\big|_{R=R_0} = 0$. The equilibrium separation R_0 (exercise 1.4.12) is contained within the messy expression:

$$R_0^2 e^{-R_0/\rho} = \frac{\rho \alpha q^2}{4\pi\varepsilon_0 z \lambda} \tag{1.51}$$

- The equilibrium energy is therefore obtained in exercise 1.4.12 as (figure 1.47)

$$U(R_0) = -N\frac{\alpha q^2}{4\pi\varepsilon_0 R_0}\left[1 - \frac{\rho}{R_0} \right] \tag{1.52}$$

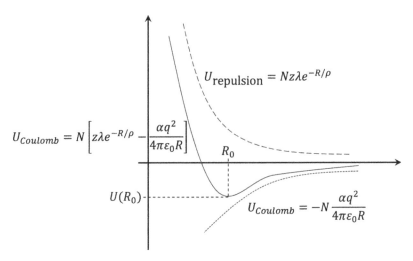

$$U_{Coulomb} = N\left[z\lambda e^{-R/\rho} - \frac{\alpha q^2}{4\pi\varepsilon_0 R}\right]$$

$$U_{repulsion} = Nz\lambda e^{-R/\rho}$$

$$U_{Coulomb} = -N\frac{\alpha q^2}{4\pi\varepsilon_0 R}$$

Figure 1.47. The Madelung energy in an ionic crystal.

1.3.3 Covalent crystals

- Covalent bonding is complicated, so we will only make a brief mention of it. From basic chemistry, we know that in covalent bonding, electrons are shared between atoms and the attractive force is the Coulomb force between the two nuclei and the shared electrons. As the shared electrons are highly localised, the covalent bond is directional.
- It is this directional nature of the bond that makes it impossible to produce simple models of cohesion in covalent crystals. The way to begin tackling this is to start with molecular physics.
- Note that ionic bonding and covalent bonding are two extremes of the transfer and sharing of electrons. Many molecules have chemical bonds that are somewhere in between.

1.3.4 Metals

- We will only briefly discuss the case of metals, as they will be discussed in greater detail in chapter 3. From basic chemistry, we recall the metallic bond as the binding between the sea of mobile electrons and the regularly arranged fixed positive ions.
- The simplest model for metallic bonding could be set up as a sea of negative electrons against a uniform background of fixed positive ions. This turns out to be quite a limited model, as the electrons need to be treated quantum mechanically and the positive ions' vibrations need to be included.

1.4 Exercises

1.4.1 Question: the Wigner–Seitz primitive cell

(Solution in section 6.1.1). For the 2D hexagonal lattice shown below, construct a Wigner–Seitz primitive cell (figure 1.48).

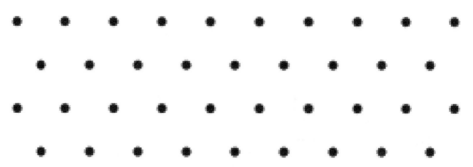

Figure 1.48. The two-dimensional hexagonal lattice for question 1.4.1.

1.4.2 Question: Miller indices for planes in 2D

(Solution in section 6.1.2.) For the 2D rectangular lattice shown (see figure 1.49), draw and label the planes (13), (01), and (23). It will suffice to draw one plane for each.

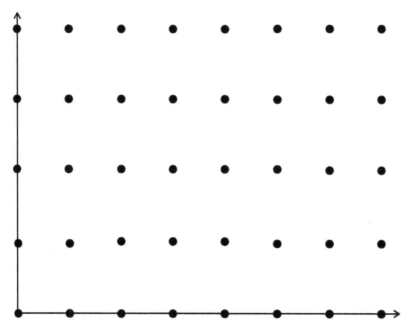

Figure 1.49. The two-dimensional rectangular lattice for question 1.4.2.

1.4.3 Question: 3D directions

(Solution in section 6.1.3.) For a 3D cubic structure, we display a unit cube below (figure 1.50). Draw the directions [101], [012], and [221] within the cube. Label the axes as well.

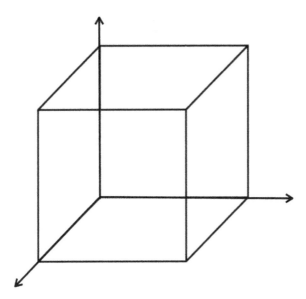

Figure 1.50. The three-dimensional unit cube for question 1.4.3.

1.4.4 Question: reciprocal space primitive translation vectors

(Solution in section 6.1.4.) Recall the example shown in figure 1.9. You now need to check certain calculations shown in the example:

1. For the real space primitive vectors, $\vec{a}_1 = a\hat{x}$ and $\vec{a}_2 = b\hat{y}$, calculate and show that the reciprocal space primitive vectors are $\vec{b}_1 = \frac{2\pi}{a}\hat{x}$ and $\vec{b}_2 = \frac{2\pi}{b}\hat{y}$.
2. Write down the explicit expression for the shortest distance between the (23) planes.

1.4.5 Question: graphene as an example of a non-Bravais lattice

(Solution in section 6.1.5.) For the 2D honeycomb lattice shown (figure 1.51), we can treat the underlying Bravais lattice as the oblique lattice. Assuming that the carbon–carbon bond length is a,

1. Write down the displacement vectors of the two carbon atoms (with respect to the oblique lattice points) in terms of a.
2. Calculate the reciprocal space primitive translation vectors for the oblique Bravais lattice in terms of a.

1.4.6 Question: hexagonal close packing

(Solution in section 6.1.6.) Show that if HCP has the same packing fraction as CCP, which is 74% or $\frac{\pi\sqrt{2}}{6}$, then we get $\frac{c}{a} = \sqrt{\frac{8}{3}}$.

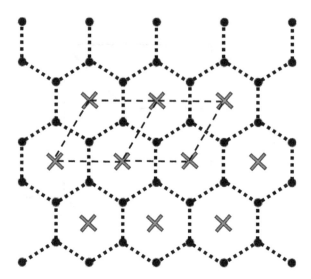

Figure 1.51. The two-dimensional honeycomb lattice for question 1.4.5.

1.4.7 Question: the Ewald sphere

(Solution in section 6.1.7.) Consider a 2D square lattice whose lattice spacing is a. An x-ray of wavelength $\lambda = \frac{2a}{5}$ is incident on the 2D crystal in the [10] direction. Note that for this question, you need to produce a scale drawing to achieve reasonable accuracy. Also note that it is acceptable to tackle the entire problem algebraically (without drawing any diagrams).

1. Draw the real space and then draw the reciprocal space. Write down their primitive translation vectors.
2. In reciprocal space, draw the Ewald sphere.
3. Note that there are five Bragg diffraction peaks. Determine their Bragg angles and write down the planes responsible for them.

1.4.8 Question: the BCC geometrical structure factor

(Solution in section 6.1.8.) Recall the discussion following equation (1.28). The zero BCC geometrical structure factor results in many points in the SC reciprocal space that can be effectively removed. Show that after removing these points from the SC reciprocal space, you get the FCC structure for the reciprocal space. This is actually sensible, as you will recall that the reciprocal space of BCC is FCC. (Hint: It will be sufficient to work with a 2D projected version of reciprocal space.)

1.4.9 Question: the FCC geometrical structure factor

(Solution in section 6.1.9.) Assume that FCC is SC with a four-atom basis.

1. Work out the geometrical structure factor for FCC. Show that you get equation (1.29).

2. Now, using a similar process to that applied for BCC, remove the points in SC reciprocal space whose geometrical structure factor is zero. Show that you get the BCC structure for the reciprocal space. This is actually sensible, as you will recall that the reciprocal space of FCC is BCC. (Hint: It will be sufficient to work with a 2D projected version of reciprocal space.)

1.4.10 Question: the Fourier series of a periodic function

(Solution in section 6.1.10.) Suppose there is a function periodic in the Bravais lattice: $f(\vec{r}) = f(\vec{r} + \vec{R})$, where $\vec{R}$ is the real space translation vector. Show that in the Fourier series $f(\vec{r}) = \sum_{\vec{k}} f(\vec{k}) e^{i\vec{k}\cdot\vec{r}}$, the $\vec{k}$s actually have to be the reciprocal translation vectors $\vec{G}$.

This means that the correct form of the Fourier series for such a function is $f(\vec{r}) = \sum_{\vec{G}} f_{\vec{G}} e^{i\vec{G}\cdot\vec{r}}$, as we learnt above. (Hint: make use of this mathematical result: $\frac{1}{V} \int d^3\vec{r} \; e^{i(\vec{k}-\vec{k}')\cdot\vec{r}} = \delta_{\vec{k}\vec{k}'}$.)

1.4.11 Question: the Lennard-Jones potential

(Solution in section 6.1.11.) For the Lennard-Jones potential $\phi(r)$ that models the binding in rare-gas solids, the total potential energy in the solid is

$$U = 2N\varepsilon \left[A_{12} \left(\frac{\sigma}{r} \right)^{12} - A_6 \left(\frac{\sigma}{r} \right)^6 \right]. \tag{1.53}$$

1. Calculate and show that the equilibrium separation is $r_0 = \left(\frac{2A_{12}}{A_6} \right)^{1/6} \sigma$.
2. Calculate and show that the cohesive energy (energy at r_0) of the solid is $U = -\frac{N\varepsilon A_6^2}{2A_{12}}$.
3. Sketch the Lennard-Jones potential and label the equilibrium separation and the cohesive energy in your sketch.

1.4.12 Question: ionic bonding

(Solution in section 6.1.12.) For this model of total energy in an ionic solid,

$$U = N \left[z\lambda e^{-R/\rho} - \alpha \frac{q^2}{4\pi\varepsilon_0 R} \right] \tag{1.54}$$

where the symbols are defined right below equation (1.47).
1. Show that solving for the equilibrium separation results in the complicated expression $R_0^2 e^{-R_0/\rho} = \frac{\rho \alpha q^2}{4\pi\varepsilon_0 z\lambda}$.
2. Then show that the cohesive energy can be obtained as $U(R_0) = -N\frac{\alpha q^2}{4\pi\varepsilon_0 R_0} \left[1 - \frac{\rho}{R_0} \right]$.

1.4.13 Question: ionic bonding with a different model for repulsion

(Solution in section 6.1.13.) We now model the repulsive force (in the ionic solid) as a power form $\frac{1}{R^n}$, where we assume that n is an integer. The total lattice energy $U(R)$ in terms of this repulsive model is then

$$U(R) = N\left[z\lambda\frac{1}{R^n} - \alpha\frac{q^2}{4\pi\varepsilon_0 R}\right] \tag{1.55}$$

1. Show that the cohesive energy can be derived as $U(R_0) = -N\frac{\alpha q^2}{4\pi\varepsilon_0 R_0}\left(1 - \frac{1}{n}\right)$.
2. Suppose NaCl has a bulk modulus of $B = 2.4 \times 10^{11}$ dyne cm^{-2}. Show that the integer n is approximately eight. Note that for NaCl, $R_0 = 2.820$ Å and $\alpha = 1.747$.

1.5 Appendix: discrete symmetry operations

Throughout chapter 1, we have been using words and phrases like 'symmetry' and 'related by symmetries'. This appendix very briefly takes a deeper look at the mathematics of symmetry in solids.

- The mathematical language for symmetry is group theory. This is because group theory deals with operations that leave certain objects unchanged. These operations are symmetry operations.
- For solids, the first symmetry operation is (discrete) translation. There are then three more fundamental conversion operations:
  - Rotations about axes through the origin
  - Reflections in planes containing the origin
  - Inversion, which changes $\vec{r}$ into $-\vec{r}$
- The group that a solid transforms under is called the 'space group'. If the translational part of the space group is set to zero, we get the 'point group'.
- In 2D, there are ten point groups consistent with translational symmetry and 17 space groups (space groups in 2D are sometimes called plane groups). We are not going to say more about 2D in this appendix.
- In 3D, there are 32 point groups consistent with translational symmetry and 230 space groups. We are not going to derive any of these. We just want to highlight how they are obtained:

  Step 1: We list the various fundamental symmetry operations in crystallographic point groups (in Schoenflies notation):
  1. E: identity.
  2. C_n: rotation through $\frac{2\pi}{n}$. For solids, n can only take the values one, two, three, four, and six. This is because fivefold symmetry is not compatible with translation.
  3. σ: reflection in a plane.
  4. σ_h: reflection in the 'horizontal' plane, i.e. the plane through the origin perpendicular to the axis of highest rotational symmetry.

Table 1.2. The seven crystal systems in 3D are defined by the symmetries shown. The table also shows how the 32 point groups are distributed over the seven crystal systems. The table also shows how the 230 space groups are distributed over the seven crystal systems. It is typically stated that the crystal systems are defined by the relationships between lattice parameters; however, this is incorrect. They are defined by symmetries.

Crystal system	Defining symmetry	Number of point groups	Number of space groups
Triclinic	none	2	2
Monoclinic	One mirror plane or one twofold rotation	3	13
Orthorhombic	Each axis has either a mirror plane or a twofold rotation axis or both	3	59
Hexagonal	One sixfold rotation axis	7	45
Tetragonal	One fourfold rotation axis	7	68
Trigonal or rhombohedral	One threefold rotation axis	5	7
Cubic	Four intersecting threefold rotation axes	5	36

5. σ_v: reflection in a 'vertical plane', i.e. one passing through the axis of highest symmetry.
6. σ_d: reflection in a 'diagonal' plane, i.e. one containing the symmetry axis and bisecting the angle between the twofold axes perpendicular to the symmetry axis. This is a special case of σ_v.
7. S_n: improper rotation through $\frac{2\pi}{n}$. Improper rotation is a combination of rotation C_n and inversion.

Step 2: The 32 point groups are then enumerated by combining various fundamental symmetry operations together.

Step 3: The 230 space groups are then enumerated from the point groups by introducing translation. In some cases, it is not just simple translation; these operations are called glide planes and screw axes. Space groups with glide planes and/or screw axes are called 'symmorphic'; otherwise, they are called 'non-symmorphic'. Out of the 230 space groups, 73 are symmorphic and 157 are non-symmorphic.

• Finally, when we say that 'planes (in the cubic crystal system) are equivalent by symmetry', it means that all the planes related by the defining symmetry operation of the cubic system: 'four intersecting threefold rotation axes' are equivalent (table 1.2).

References

[1] Silva-Ramírez E-L, Cumbrera-Conde I, Cano-Crespo R and Cumbrera F-L 2023 Machine learning techniques for the *ab initio* Bravais lattice determination *Expert Syst.* **40** e13160
[2] Kittel C 2005 *Introduction to Solid State Physics* 8th edn (New York: Wiley) https://www.wiley.com/en-us/Introduction+to+Solid+State+Physics%2C+8th+Edition-p-9780471415268
[3] Alloul H 2010 *Introduction to the Physics of Electrons in Solids* (Berlin: Springer)

[4] Novakovic R 2006 Hexagonal Close-Packed for Iron (http://rodolfo-novakovic.blogspot.com/2006/08/hexagonal-close-packed-for-iron.html)
[5] Ashcroft N W and Mermin N D 2022 *Solid State Physics* (Boston, MA: Cengage Learning)
[6] Sólyom J 2007 *Fundamentals of the Physics of Solids: Volume 1: Structure and Dynamics* (Berlin: Springer)
[7] Patterson J D and Bailey B C 2018 *Solid-State Physics: Introduction to the Theory* 3rd edn (Berlin: Springer)
[8] Quinn J J and Yi K-S 2018 *Solid State Physics: Principles and Modern Applications* 2nd edn (Berlin: Springer)

IOP Publishing

Crystalline Solid State Physics
An interactive guide
Meng Lee Leek

Chapter 2

Lattice vibrations and thermodynamics

In this chapter, we shall remove the assumption we made in chapter 1, where we said that the atoms are in fixed positions. So now the atoms can make small vibrations, which has important consequences.

The first consequence is that waves can form in the lattice (i.e. lattice waves). Since these waves are microscopic, the quantum mechanical description of the waves is important.

The second consequence is that since atoms can vibrate, they have degrees of freedom to store heat. This means that the lattice has a heat capacity. It turns out that experimental measurements of the heat capacity show that a quantum mechanical understanding of lattice waves is essential.

2.1 Lattice vibrations

2.1.1 Harmonic approximation

- Let us consolidate some facts we learnt in chapter 1.
  - We started by considering a lattice; this means we assumed that the atoms are fixed and do not move (and are regularly spaced with translational invariance).
  - We then realised that the potential energy between a pair of atoms always has an 'L' shape with a dip; the minimum point is where the crystal is stable and atoms prefer to be in their equilibrium separation. Note that this analysis was done at $T = 0$.
- Thus, at $T \neq 0$, the atoms should wiggle a little about the minimum potential energy point.
- It follows that the potential energy of the wiggling atoms is a 'little smiley face' near the minimum potential energy point. We then approximate the 'little smiley face' about the minimum potential energy point as a quadratic/ harmonic potential. This is called the harmonic approximation (figure 2.1).
- We now want to relate this spring constant k to the crystal potential $U(\vec{R})$.

doi:10.1088/978-0-7503-5217-8ch2

© IOP Publishing Ltd 2023

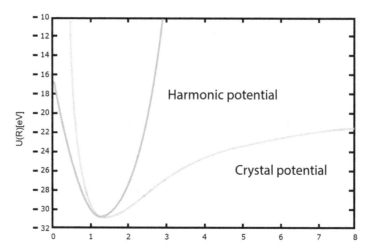

Figure 2.1. A graphical visualisation of the harmonic approximation. The quadratic/harmonic potential energy is $U = \frac{1}{2}kx^2$, where x is a small displacement and k is the 'spring constant'. So at $T \neq 0$, we shall picture atoms as being held by springs and as wiggling.

crystal potential $= U(\vec{R})$

| write the separation vector as $\vec{R} = \vec{R}_0 + \vec{u}$ (2.1)

| where $\vec{u}$ is a small displacement vector

$U(\vec{R}) = U(\vec{R}_0 + \vec{u})$

| recall Taylor series $f(x) = f(a) + \left.\dfrac{df}{dx}\right|_{x=a} (x - a)$

$\left| + \dfrac{1}{2!} \dfrac{d^2 f}{dx^2}\right|_{x=a} (x - a)^2 + \cdots$ (2.2)

| now $x \Longrightarrow R$ and $a \Longrightarrow R_0$ where we also assume 1D

$U(R) = U(R_0) + \left.\dfrac{\partial U}{\partial R}\right|_{R=R_0} u + \left.\dfrac{1}{2} \dfrac{\partial^2 U}{\partial R^2}\right|_{R=R_0} u^2 + \cdots$

| first term: $U(R_0)$ is just some number

| second term: $\left.\dfrac{\partial U}{\partial R}\right|_{R=R_0} = 0$ because R_0 is a stationary point   (2.3)

| third term: quadratic term and is what we want!

$$U(R) = \left.\dfrac{1}{2} \dfrac{\partial^2 U}{\partial R^2}\right|_{R=R_0} u^2 + \cdots \qquad (2.4)$$

Comparing with $U = \frac{1}{2}kx^2$, we find that the spring constant is $k = \left.\dfrac{\partial^2 U}{\partial R^2}\right|_{R=R_0}$.

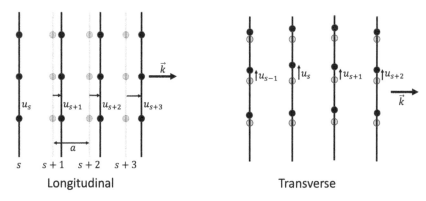

Figure 2.2. Longitudinal polarisation means that the atoms vibrate parallel to the direction of propagation. Transverse polarisation means that the atoms vibrate perpendicular to the direction of propagation. Note that in a regular vibration[1], the atoms in the same plane vibrate in the same manner. This actually reduces the 3D problem to 1D! Adapted with permission from [1].

- Finally, we just want to mention that there are two polarisations for vibrating atoms in a lattice: longitudinal and transverse (see figure 2.2).

2.1.2 1D chain (monoatomic basis)

We are now going to solve the simplest example of a vibrating lattice problem, namely a 1D solid with a one-atom basis that is vibrating in 1D, i.e. the atoms are only vibrating left and right. This also implies that the vibrations are longitudinal.

By 'solving', we mean we are going to find the dispersion relation. The dispersion relation is the relation between the (angular) frequency ω and the wave vector $q = \frac{2\pi}{\lambda}$ of the vibration.

We start the derivation of the dispersion relation with Newton's second law. We use the sth atom as a reference (figure 2.3).

$$Ma = F_{\text{net}}$$
$$\quad | \quad \text{recall Hooke's law: } F = kx \text{ where } x \text{ is the displacement} \tag{2.5}$$

$$M\frac{d^2u_s}{dt^2} = k(u_{s-1} - u_s) + k(u_{s+1} - u_s) \tag{2.6}$$

$$= k(u_{s+1} + u_{s-1} - 2u_s)$$
$\quad |$ assume a wave solution $u_s = ue^{i(qx_s - \omega t)}$ where the complex form is used
$\quad |$ note that $x_s = sa$, the equilibrium position of sth atom so $u_s = ue^{i(qsa - \omega t)}$ (2.7)
$\quad |$ substitute the solution into both sides of the equation

---

[1] The proper phrase is 'normal mode vibrations' from classical mechanics.

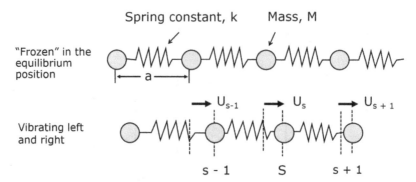

Figure 2.3. The setup for the 1D monoatomic basis dispersion derivation.

$$-M\omega^2 u e^{i(qsa-\omega t)} = ku(e^{i(q(s+1)a-\omega t)} + e^{i(q(s-1)a-\omega t)} - 2e^{i(qsa-\omega t)}) \tag{2.8}$$

$$= ku(e^{iqa} + e^{-iqa} - 2)e^{i(qsa-\omega t)}$$

| cancel the factor $e^{i(qsa-\omega t)}$ and make ω^2 the subject

$$\tag{2.9}$$

$$\omega^2 = \frac{k}{M}(2 - 2\cos qa)$$

| use identity: $\cos qa = 1 - 2\sin^2\frac{qa}{2}$

$$\tag{2.10}$$

$$\boxed{\omega = \sqrt{\frac{4k}{M}}\left|\sin\frac{qa}{2}\right|} \tag{2.11}$$

We have finished, and this is the dispersion relation of a 1D chain (with 1D vibrations) with a one-atom basis. We can now discuss its physics.

- Because we claimed in chapter 1 that the first Brillouin Zone (1BZ) is very important as it contains all the physics (the proof will be given in chapter 3, section 3.2.1.2), we shall sketch the dispersion relation in the 1BZ. The real-space translation vector is $\vec{R} = n_1\vec{a}_1$, where $\vec{a}_1 = a\hat{x}$. The reciprocal space primitive vector is then $\vec{b}_1 = \frac{2\pi}{a}\hat{x}$ and by the Wigner–Seitz construction, the 1BZ is $q \in \left\{-\frac{\pi}{a}, \frac{\pi}{a}\right\}$. A sketch of the dispersion in the 1BZ is shown in figure 2.4.
- The importance of the 1BZ is that it contains all the physics. Outside the 1BZ, the physics is just repeated. The proof is given in chapter 3, page 94 but figure 2.5 gives a clue to this.
- We introduce two velocities that characterise waves:
  - **Phase velocity, v_p:** this is the velocity of a single 'pure wave'. This is the familiar $v_p = f\lambda = \frac{\omega}{q}$.
  - **Group velocity, v_g:** this is the velocity of a 'packet of waves' which is made up of many 'pure waves'. It is tricky to define a velocity of a wave packet, since it is made up of waves of various velocities. The sensible

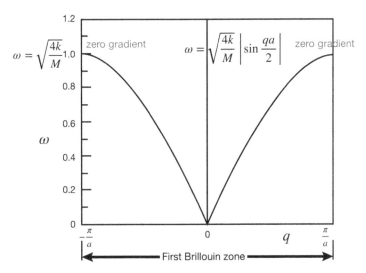

Figure 2.4. The dispersion relation for a (1D) vibrating 1D chain with a monoatomic basis in the 1BZ. The value at which ω intercepts the ω-axis is $\sqrt{\frac{4k}{M}}$. It is important to note that the gradient of the dispersion is zero at the 1BZ boundaries ($q = -\frac{\pi}{a}$ and $q = \frac{\pi}{a}$). Adapted with permission from [1].

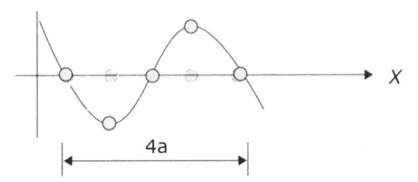

Figure 2.5. Consider two lattice waves shown as a solid line and a dashed line. The solid line has $\lambda = 4a$ and so $q = \frac{2\pi}{\lambda} = \frac{1}{2}\frac{\pi}{a}$, which is inside the 1BZ. The dashed line has $\lambda = \frac{4}{5}a$ and so $q = \frac{2\pi}{\lambda} = \frac{5}{2}\frac{\pi}{a}$ which is outside the 1BZ. Considering the fact that we only see the atoms and not any of the lines, these two waves look the same— thus, they are physically the same! This gives the idea that there is repeated physics outside the 1BZ.

way is to define the velocity of the 'envelope' of the wave packet which is $v_g = \frac{d\omega}{dq}$ which is the gradient of the dispersion relation. This envelope carries the energy of the wave. For this dispersion, the group velocity is

$$v_g = \frac{d}{dq}\sqrt{\frac{4k}{M}}\sin\frac{qa}{2} = a\sqrt{\frac{k}{M}}\cos\frac{qa}{2}.$$

- **Long-wavelength limit** ($q \to 0$): the long-wavelength limit is actually interesting because it represents the lowest-energy ($E = \hbar\omega$) region, which therefore contains the first frequencies to get excited as the temperature rises from $T = 0$. Recall that for small x, $\sin x \approx x$, so for small q, $\omega \approx \sqrt{\frac{4k}{M}}\frac{qa}{2} = qa\sqrt{\frac{k}{M}}$.

Thus, in this limit, $\omega \propto q$, which is a linear relationship. And in this limit,

$$v_g = a\sqrt{\frac{k}{M}}.$$

2.1.3 A one-dimensional chain (diatomic basis)

The second simplest example of the vibrating lattice problem, which we will now solve, is the 1D solid with a two-atom basis. The vibration will still be 1D, i.e. the atoms are only vibrating left and right. Again, we want to obtain the dispersion relation starting with Newton's second Law.

Recall from figure 2.2 that the 1D problem can be justified by looking at vibrating planes in a 3D solid. Here, we can justify the same thing by saying that each vibrating plane contains one type of atom only. For example, in the CsCl structure, the lattice wave in the [100] direction is made up of one plane of Cl ions and another plane of Cs ions, which is just like a 1D solid with a two-atom basis (figure 2.6).

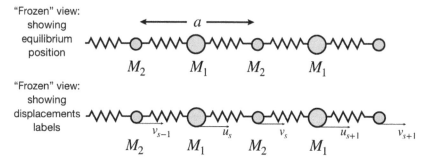

Figure 2.6. The setup for the 1D diatomic basis dispersion derivation.

We proceed with Newton's second law for M_1 (left column) and M_2 (right column).

$$M_1\frac{d^2u_s}{dt^2} = k(v_s + v_{s-1} - 2u_s) \qquad\qquad M_2\frac{d^2v_s}{dt^2} = k(u_{s+1} + u_s - 2v_s)$$

| insert solution: $u_s = ue^{iqsa-i\omega t}$ into LHS | insert solution: $v_s = ve^{iqsa-i\omega t}$ into LHS

$$-\omega^2 M_1 u_s = k(v_s + v_{s-1} - 2u_s) \qquad\qquad -\omega^2 M_2 v_s = k(u_{s+1} + u_s - 2v_s)$$

| insert all solutions then cancel $e^{iqsa-i\omega t}$ | insert all solutions then cancel $e^{iqsa-i\omega t}$

$$-\omega^2 M_1 u = kv(1 + e^{-iqa}) - 2ku \qquad\qquad -\omega^2 M_2 v = ku(e^{iqa} + 1) - 2kv$$

We write the coupled equations in a matrix form:

$$\begin{pmatrix} 2k - M_1\omega^2 & -k(1 + e^{-iqa}) \\ -k(e^{iqa} + 1) & 2k - M_2\omega^2 \end{pmatrix}\begin{pmatrix} u \\ v \end{pmatrix} = 0 \tag{2.12}$$

A solution exists when the determinant of the coefficient matrix $= 0 \,|$

$$\det \begin{vmatrix} 2k - M_1\omega^2 & -k(1 + e^{-iqa}) \\ -k(e^{iqa} + 1) & 2k - M_2\omega^2 \end{vmatrix} = 0 \tag{2.13}$$

expanding out the determinant gives $|$

$$(2k - M_1\omega^2)(2k - M_2\omega^2) - k^2(1 + e^{-iqa})(1 + e^{iqa}) = 0 \tag{2.14}$$
$$\text{use } e^{iqa} + e^{-iqa} = 2\cos qa \,\,|$$

$$M_1M_2\omega^4 - 2k(M_1 + M_2)\omega^2 + 2k^2(1 - \cos qa) = 0 \tag{2.15}$$
$$\text{now use } \cos qa = 1 - 2\sin^2\left(\frac{qa}{2}\right) \,|$$

$$M_1M_2\omega^4 - 2k(M_1 + M_2)\omega^2 + 4k^2\sin^2\left(\frac{qa}{2}\right) \tag{2.16}$$

solving this quadratic equation gives $|$

$$\boxed{\omega^2 = k\left(\frac{1}{M_1} + \frac{1}{M_2}\right) \pm k\sqrt{\left(\frac{1}{M_1} + \frac{1}{M_2}\right)^2 - \frac{4}{M_1M_2}\sin^2\left(\frac{qa}{2}\right)}} \tag{2.17}$$

The first surprise is that are there two solutions for ω^2. It turns out that because of the two-atom basis, there are two ways in which a primitive cell can vibrate: the acoustic branch and the optical branch[2] (figure 2.7).

- **Acoustic branch:** this is the kind of vibration in which the 'centre of mass' of the unit cell moves significantly. See figure 2.8.
- **Optical branch:** this is the kind of vibration in which the 'centre of mass' of the unit cell hardly moves. See figure 2.8.
- **Long-wavelength limit:** again, we use $\sin x \approx x$ but we now have two cases:
  - For the optical branch (we can simply put $q = 0$):

$$\omega^2 = k\left(\frac{1}{M_1} + \frac{1}{M_2}\right) + k\sqrt{\left(\frac{1}{M_1} + \frac{1}{M_2}\right)^2 - \frac{4}{M_1M_2}\sin^2\left(\frac{qa}{2}\right)} \tag{2.18}$$

---

[2] In this book, I merely state and explain the two branches. I do not derive how the atoms vibrate, as stated in my description. The word 'acoustic' indeed has something to do with sound. The acoustic branch is the one with the higher group velocity and, in the long-wavelength limit, is the propagation of sound in the solid.

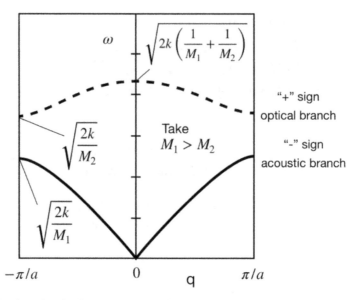

Figure 2.7. The dispersion (in the 1BZ) of a vibrating 1D chain with a diatomic basis. The positive sign solution is the optical branch solution. The negative sign solution is the acoustic branch solution. Note that the monoatomic basis only has the acoustic branch solution.

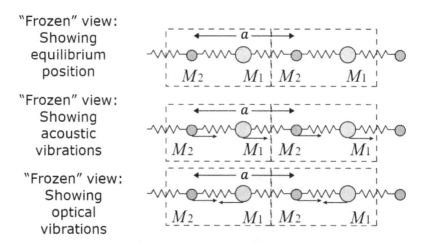

Figure 2.8. An illustration of the approximate movements of the two-atom basis in different vibrational modes. Imagine the 'centre of mass' of each unit cell and imagine how much the 'centre of mass' moves under each type of vibration. For the middle diagram, both atoms move in the same direction, thus the centre of mass moves quite a bit, while for the bottom diagram, the atoms move oppositely, thus the centre of mass moves only a little.

$$\omega^2 \approx 2k\left(\frac{1}{M_1} + \frac{1}{M_2}\right) \tag{2.19}$$

which is a constant.

– For the acoustic branch (we cannot simply put $q = 0$, otherwise we get $\omega^2 = 0$):

$$\omega^2 = k\left(\frac{1}{M_1} + \frac{1}{M_2}\right) - k\sqrt{\left(\frac{1}{M_1} + \frac{1}{M_2}\right)^2 - \frac{4}{M_1 M_2}\sin^2\left(\frac{qa}{2}\right)} \qquad (2.20)$$

$$\omega^2 \approx k\left(\frac{1}{M_1} + \frac{1}{M_2}\right) - k\sqrt{\left(\frac{1}{M_1} + \frac{1}{M_2}\right)^2 - \frac{4}{M_1 M_2}\left(\frac{qa}{2}\right)^2} \qquad (2.21)$$

$$\omega^2 = k\left(\frac{1}{M_1} + \frac{1}{M_2}\right) - k\left(\frac{1}{M_1} + \frac{1}{M_2}\right)\sqrt{1 - \frac{4}{M_1 M_2}\left(\frac{qa}{2}\right)^2\left(\frac{1}{M_1} + \frac{1}{M_2}\right)^{-2}}$$

| use binomial expansion: $\sqrt{1 - x} \approx 1 - \frac{1}{2}x$ \qquad (2.22)

$$\omega^2 = k\left(\frac{1}{M_1} + \frac{1}{M_2}\right) - k\left(\frac{1}{M_1} + \frac{1}{M_2}\right)\left(1 - \frac{1}{2}\frac{4}{M_1 M_2}\left(\frac{qa}{2}\right)^2\left(\frac{1}{M_1} + \frac{1}{M_2}\right)^{-2}\right)$$

$$\omega^2 = \frac{\frac{1}{2}k}{M_1 + M_2}q^2 a^2 \qquad (2.23)$$

which is again $\omega \propto q$, a linear relation, just as in the monoatomic case.

2.1.4 Lattice vibrations in 3D

- Let us recall: for a 1D solid with 1D vibrations and a monoatomic basis, we have only one acoustic branch. As this vibration is longitudinal, it should be called the longitudinal–acoustic (LA) branch.
- Let us recall: for a 1D solid with 1D vibrations and a diatomic basis, we have one acoustic branch (LA) and one optical branch (longitudinal optical (LO)).
- Let us infer: for a 1D solid with 3D vibrations and a monoatomic basis, we should have three acoustic branches, as there are three directions in which the atoms can vibrate. The two new directions are transverse (T). The three branches are therefore one LA branch and two transverse acoustic (TA) branches.
- Let us infer that for a 1D solid with 1D vibrations and a p-atom basis, we should have 1 LA branch and $(p - 1)$ optical branches. The reason for this is that acoustic vibrations shift the 'centre of mass' of the unit cell significantly but in the case of 1D vibrations, there is only one way in which vibration can occur, which is that all the p atoms vibrate together in the same (longitudinal) direction.
- Let us infer the most general case, namely that of a 3D solid with 3D vibrations and a p-atom basis; we should now get three acoustic branches

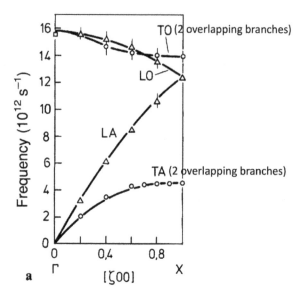

Figure 2.9. The actual dispersion relation of lattice waves of a diamond structure in the [100] direction. The diamond structure is FCC with a two-atom basis. Indeed, there are $3(2 - 1) = 3$ optical branches (1 LO and two TO) and the usual three acoustic branches. Reprinted by permission from Springer Nature [2], Copyright (2009).

(one LA branch, two TA branches) and $3(p - 1)$ optical branches (some are LO branches and some others are TO branches) (figure 2.9).

2.1.5 The quantum story of lattice vibrations: phonons

Everything that was discussed earlier is classical: atoms in the solid vibrate and form waves in the lattice. However, at this small scale, quantum effects are non-negligible, so we will briefly state the quantum version of lattice vibrations.

- The quantum discussion always starts with the Hamiltonian operator. The Hamiltonian is made of the kinetic energy of the atoms and the harmonic potential energy between the atoms. It turns out that (not shown here) the Hamiltonian operator can be rewritten into a harmonic oscillator form summed over the normal modes of vibration with the normal mode frequencies labelled by ω_l for the lth mode.

- The harmonic oscillator Hamiltonian has energy eigenvalues $E = \hbar\omega\left(n + \frac{1}{2}\right)$, where n is an integer. Similarly, the (rewritten) lattice vibration Hamiltonian has energy eigenvalues summed over all the modes $E = \sum_l \hbar\omega_l\left(n_l + \frac{1}{2}\right)$.

- A basic quantum ('particle') of lattice vibrations is called a phonon. Note that n_l can be described as n phonons in the lth normal mode. This is, in some sense, the wave–particle duality.

- It turns out that the quantum nature of lattice vibrations shows up in experimental measurements of the lattice's heat capacity at low temperatures. We will discuss this in detail in section 2.2.2.

2.1.6 Interactions and scattering

Since quanta of lattice vibrations can now be treated as particles called phonons, we briefly state the rules used to treat phonons as particles and describe how they can interact.

- The phonon, as a particle, can be assigned 'momentum' $\hbar\vec{q}$, and we can talk about conservation of 'momentum' when there is interaction/scattering. However, as mentioned earlier, the 1BZ contains all the physics and outside the 1BZ are repetitions, so the word 'momentum' is misleading. We shall call it the conservation law of wave vectors instead. The conservation law for wave vectors is also known as selection rules.
- Note again that $\vec{q} \in 1BZ$ contains all the physics, and any wave vector outside the 1BZ is unnecessary. We shall therefore restrict all wave vectors to the 1BZ by adding the reciprocal translation vector $\vec{G}$ if needed, to map $\vec{q}$ back to the 1BZ.
- Taking the above facts into account, we can now write the selection rules explicitly for electron–phonon interactions. We let $\vec{k}'$ denote the final wave vector of the electron after interaction, $\vec{k}$ denote the initial wave vector of the electron, $\vec{q}$ denote the wave vector of the phonon and $\vec{G}$ denote the reciprocal translation vector needed to bring $\vec{k}'$ into the 1BZ (if necessary).
  - The selection rule for interactions involving phonon creation is: $\vec{k}' = \vec{k} - \vec{q} + \vec{G}$
  - The selection rule for interactions involving phonon absorption is: $\vec{k}' = \vec{k} + \vec{q} + \vec{G}$

2.2 Lattice thermodynamics

2.2.1 Experimental results for lattice heat capacity

The lattice atoms can vibrate, thus these degrees of freedom can be used to store heat. This is the heat capacity of the lattice.

- In thermodynamics, the definitions of heat capacities are:

$$\text{constant pressure: } c_P = \left(\frac{\partial U}{\partial T}\right)_P, \quad \text{constant volume: } c_V = \left(\frac{\partial U}{\partial T}\right)_V \qquad (2.24)$$

  where U is the internal energy of the system.
- Note that the internal energy U is typically calculated using techniques taken from statistical mechanics.
- The quantum statistical mechanics calculation of U and the classical statistical mechanics calculation of U are different! Essentially, this is the crucial reason why the quantum explanation works!
- Experiments usually measure c_P, while theoretical calculations give c_V more easily. The two heat capacities are related (as taught in thermal physics) but we do not really need to worry about it here, as $c_P \approx c_V$ for solids.

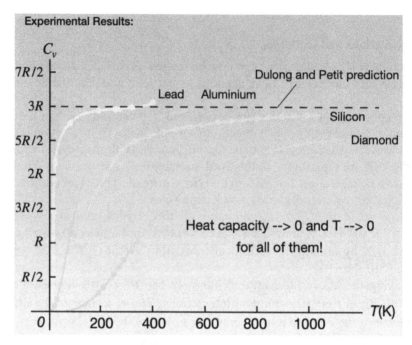

Figure 2.10. Experimental measurements of the heat capacities of metals and insulators. All show reasonable agreement with the Dulong–Petit law (see section 2.2.2) at high temperatures. At low temperatures, the heat capacities go to zero. At low temperatures, the relationship for insulators is $c_V \propto T^3$ and for metals it is $c_V = \gamma T + \alpha T^3$, where γ and α are constants.

- Let us look at the experimental measurement of the lattice heat capacities of conductors and insulators (figure 2.10); we try to explain these in section 2.2.2.

2.2.2 An explanation of lattice heat capacity

2.2.2.1 The classical explanation: Dulong–Petit law
We shall avoid using classical statistical mechanics to calculate U. Instead, we will just use the equipartition theorem (which is actually the same) to derive U.

- The equipartition theorem states that each degree of freedom can store $\frac{1}{2}k_BT$ of energy (at thermal equilibrium).
- A 1D harmonic oscillator has energy $\frac{1}{2}mv^2 + \frac{1}{2}kx^2$, which means two degrees of freedom: one of kinetic energy and one of potential energy; thus, it can store $\frac{1}{2}k_BT + \frac{1}{2}k_BT = k_BT$ of energy. So a 3D harmonic oscillator can store $3k_BT$ of energy.
- As mentioned in section 2.1.5, the Hamiltonian of the 3D vibrating lattice can be written as a sum over normal modes. Each normal mode is like a harmonic oscillator. If there are N atoms in the solid, there are N normal modes (see section 2.2.3.1). So the internal energy of the solid is $U = 3Nk_BT$.

- The heat capacity is $c_V = \left(\frac{\partial U}{\partial T}\right)_V = 3Nk_B$. This is called the Dulong–Petit law, which states that the (constant volume) specific heat capacity is just a constant!
- Thus, the classical explanation can only explain the high-temperature part of the experimental results, which is unsatisfactory. We shall see that the quantum explanation explains the entire behaviour of the specific heat capacity all the way from low temperatures to high temperatures.

2.2.2.2 The quantum explanation: Einstein's model and Debye's model
We now have to use quantum statistical mechanics to derive U, the internal energy, and then use $c_V = \left(\frac{\partial U}{\partial T}\right)_V$ to get the heat capacity. However, the quantum calculation of U is not trivial.

- Recall that the energy eigenvalue of the phonons (the quanta of lattice vibrations) is $E = \sum_l \hbar\omega_l\left(n_l + \frac{1}{2}\right)$, where l is the mode and n_l is the number of phonons in the lth mode. From the dispersion relation, we saw that the lth mode is labelled by a branch (denoted by j) and a wave vector $\vec{q}$, so we need to rewrite E as $E = \sum_j \sum_{\vec{q}} \hbar\omega_{j\vec{q}}\left(n_{j\vec{q}} + \frac{1}{2}\right)$. This E is a suitable starting point from which to calculate the internal energy U.
- Quantum statistical mechanics says that the thermal energy decides the probability of occupation of a state. The higher the energy level, the lower the occupation probability. Thus, the higher the energy of the mode, the fewer the number of phonons. This means that $n_{j\vec{q}}$ has to follow a distribution. Because phonons are bosons (not explained here), they follow the same distribution as photons (which are also bosons), which is the Planck distribution $\langle n_{j\vec{q}} \rangle = \frac{1}{e^{\hbar\omega_{j\vec{q}}/k_B T} - 1}$.
- So we can write the quantum internal energy of the solid as

$$U = \sum_j \sum_{\vec{q}} \hbar\omega_{j\vec{q}}\left(\langle n_{j\vec{q}} \rangle + \frac{1}{2}\right) = \sum_j \sum_{\vec{q}} \frac{\hbar\omega_{j\vec{q}}}{e^{\hbar\omega_{j\vec{q}}/k_B T} - 1} + \sum_j \sum_{\vec{q}} \frac{1}{2}\hbar\omega_{j\vec{q}} \qquad (2.25)$$

The second term: $\sum_j \sum_{\vec{q}} \frac{1}{2}\hbar\omega_{j\vec{q}}$ is independent of temperature, so when we take the derivative $\frac{\partial}{\partial T}$ to get the heat capacity, we will get zero; therefore, let us drop this term now. The big question is: how are we supposed to calculate the double sum? This will be done via two tricks.

- **Trick 1:** we insert a delta function $\delta(\omega - \omega_{j\vec{q}})$ and integrate over the parameter ω. If we carry out the integration, the trick will be undone.

$$U = \sum_j \sum_{\vec{q}} \frac{\hbar\omega_{j\vec{q}}}{e^{\hbar\omega_{j\vec{q}}/k_B T} - 1} = \sum_j \sum_{\vec{q}} \int d\omega\, \delta(\omega - \omega_{j\vec{q}}) \frac{\hbar\omega}{e^{\hbar\omega/k_B T} - 1} \qquad (2.26)$$

|denote $\sum_{\bar{q}} \delta(\omega - \omega_{j\bar{q}}) = D_j(\omega)$

$\quad$ (2.27)

|note that $D_j(\omega)$ is called the density of states of branch j

$$U = \sum_j \int d\omega\, D_j(\omega) \frac{\hbar\omega}{e^{\hbar\omega/k_B T} - 1} \qquad (2.28)$$

The reason that the expression $D_j(\omega) = \sum_{\bar{q}} \delta(\omega - \omega_{j\bar{q}})$ is called the density of states is because the delta function 'counts' whenever ω equals the dispersion $\omega_{j\bar{q}}$. There could be more than one situation in which $\omega = \omega_{j\bar{q}}$, so $D_j(\omega)$ 'counts' the number of states with frequency ω.

- **Trick 2:** assume a simplified dispersion relation. Actually, we shall assume two simplified dispersion relations, which then give rise to Einstein's model and Debye's model.

Einstein's model: this model assumes that all modes have the same frequency ω_E and (from appendix 2.2.3) the DOS, in this model, is $D_j(\omega) = N\delta(\omega - \omega_E)$ for every branch j. Thus, the calculation of the internal energy U can now continue easily.

$$U = \sum_j \int d\omega\, D_j(\omega) \frac{\hbar\omega}{e^{\hbar\omega/k_B T} - 1}$$

| insert $D_j(\omega) = N\delta(\omega - \omega_E)$

$\quad$ (2.29)

| and carry out the integration via the delta function

$$U = \sum_j \frac{N\hbar\omega_E}{e^{\hbar\omega_E/k_B T} - 1} \qquad (2.30)$$

| assume there are three branches with the same frequency ω_E

$$U = \frac{3N\hbar\omega_E}{e^{\hbar\omega_E/k_B T} - 1} \qquad (2.31)$$

We now consider the heat capacity:

$$c_V = \left(\frac{\partial U}{\partial T}\right)_V = \frac{3N\hbar\omega_E e^{\hbar\omega_E/k_B T} \dfrac{\hbar\omega_E}{k_B T^2}}{(e^{\hbar\omega_E/k_B T} - 1)^2} \qquad (2.32)$$

$$c_V = 3Nk_B \left(\frac{\hbar\omega_E}{k_B T}\right)^2 \frac{e^{\hbar\omega_E/k_B T}}{(e^{\hbar\omega_E/k_B T} - 1)^2}. \qquad (2.33)$$

As an aside, we note that in the derivation above, we assumed that the solid has three optical branches. We recall the counting of 3D optical modes to be $3(p-1)$, where p is the number of basis atoms. Thus, by assuming three branches of optical modes, we assume there are two basis atoms (or a diatomic basis). Let us test this expression and see whether it fits the experimental evidence.

The high-temperature limit: T is large, so $\frac{\hbar \omega_E}{k_B T}$ is small; thus, we can Taylor expand $e^{\hbar \omega_E/k_B T}$.

$$c_V = 3Nk_B \left(\frac{\hbar \omega_E}{k_B T}\right)^2 \frac{e^{\hbar \omega_E/k_B T}}{(e^{\hbar \omega_E/k_B T} - 1)^2}$$

$$c_V \approx 3Nk_B \left(\frac{\hbar \omega_E}{k_B T}\right)^2 \frac{1}{(1 + \frac{\hbar \omega_E}{k_B T} - 1)^2} = 3Nk_B \tag{2.34}$$

So we do get the Dulong–Petit law at high temperatures!

The low-temperature limit: $T \rightarrow 0$ means $e^{\hbar \omega_E/k_B T}$ is very large; therefore, we can approximate $e^{\hbar \omega_E/k_B T} - 1 \approx e^{\hbar \omega_E/k_B T}$.

$$c_V = 3Nk_B \left(\frac{\hbar \omega_E}{k_B T}\right)^2 \frac{e^{\hbar \omega_E/k_B T}}{(e^{\hbar \omega_E/k_B T} - 1)^2} \tag{2.35}$$

$$c_V \approx 3Nk_B \left(\frac{\hbar \omega_E}{k_B T}\right)^2 \frac{e^{\hbar \omega_E/k_B T}}{(e^{\hbar \omega_E/k_B T})^2} \tag{2.36}$$

$$c_V = 3Nk_B \left(\frac{\hbar \omega_E}{k_B T}\right)^2 e^{-\hbar \omega_E/k_B T} \tag{2.37}$$

So when $T \rightarrow 0$, the exponential factor 'overcomes' $\frac{1}{T^2}$ and $c_V \rightarrow 0$, which agrees with experiments. However, the low-temperature expression is the weird expression $\frac{e^{-\hbar \omega_E/k_B T}}{T^2}$, which does not fit experiments. We need a better model.

Debye's model: This model assumes the dispersion relation: $\omega_{j\vec{q}} = v_j q$, which is a linear relationship in which v_j is a constant which is the group velocity of the jth branch. The DOS, in this model, (see appendix 2.2.3.7) is $D_j(\omega) = \frac{V\omega^2}{2\pi^2 v_j^3}$. The internal energy is

$$U = \sum_j \int d\omega \, D_j(\omega) \frac{\hbar \omega}{e^{\hbar \omega/k_B T} - 1} \tag{2.38}$$

$$U = \sum_j \int d\omega \, \frac{V\omega^2}{2\pi^2 v_j^3} \frac{\hbar \omega}{e^{\hbar \omega/k_B T} - 1} \tag{2.39}$$

| impose a physical upper limit on the integral (see appendix 2.2.3)

|this is called the Debye cutoff frequency $\omega_D = \left(\frac{6\pi^2 v_j^3 N}{V}\right)^{1/3}$

$$J = \sum_j \int_0^{\omega_D} d\omega \, \frac{V\omega^2}{2\pi^2 v_j^3} \frac{\hbar\omega}{e^{\hbar\omega/k_B T} - 1} \tag{2.40}$$

| for simplicity, assume there are three branches and that all have the same v

$$U = \frac{3V}{2\pi^2 v^3} \int_0^{\omega_D} d\omega \, \frac{\hbar\omega^3}{e^{\hbar\omega/k_B T} - 1} \tag{2.41}$$

$$\text{| let } x = \frac{\hbar\omega}{k_B T} \text{ and } x_D = \frac{\hbar\omega_D}{k_B T}$$

$$U = \frac{3V k_B^4 T^4}{2\pi^2 v^3 \hbar^3} \int_0^{x_D} dx \, \frac{x^3}{e^x - 1}$$

| this integral is not really analytically solvable, so we leave it hanging (2.42)

| let $\Theta_D = \dfrac{\hbar\omega_D}{k_B} = x_D T$ denote the Debye temperature

$$U = 9N k_B T \left(\frac{T}{\Theta_D}\right)^3 \int_0^{\Theta_D/T} dx \, \frac{x^3}{e^x - 1} \tag{2.43}$$

The high-temperature limit: T is large, so x is small; thus, we can Taylor expand e^x.

$$U = 9N k_B T \left(\frac{T}{\Theta_D}\right)^3 \int_0^{\Theta_D/T} dx \, \frac{x^3}{e^x - 1}$$

$$U \approx 9N k_B T \left(\frac{T}{\Theta_D}\right)^3 \int_0^{\Theta_D/T} dx \, \frac{x^3}{1 + x - 1} \tag{2.44}$$

$$U = 9N k_B T \left(\frac{T}{\Theta_D}\right)^3 \int_0^{\Theta_D/T} dx \, x^2 \tag{2.45}$$

$$U = 9N k_B T \left(\frac{T}{\Theta_D}\right)^3 \left[\frac{1}{3} x^3\right]_0^{\Theta_D/T} \tag{2.46}$$

$$U = 3N K_B T \tag{2.47}$$

The heat capacity is a simple derivative: $c_V = \left(\frac{\partial}{\partial T} 3N k_B T\right)_V = 3N k_B$, which is the Dulong–Petit law at high temperatures—an excellent result!

The low-temperature limit: $T \to 0$ means x, x_D are large; thus, $\frac{\Theta_D}{T} \to \infty$.

$$U = 9N k_B T \left(\frac{T}{\Theta_D}\right)^3 \int_0^{\Theta_D/T} dx \, \frac{x^3}{e^x - 1} \tag{2.48}$$

$$U \approx 9Nk_BT\left(\frac{T}{\Theta_D}\right)^3 \int_0^\infty dx\, \frac{x^3}{e^x - 1}$$

(2.49)

| The integral is standard: $\int_0^\infty dx\, \frac{x^3}{e^x - 1} = \frac{\pi^4}{15}$

$$U = 9Nk_BT\left(\frac{T}{\Theta_D}\right)^3 \frac{\pi^4}{15}$$

(2.50)

$$U = \frac{3\pi^4 Nk_BT^4}{5\Theta_D^3}$$

(2.51)

The heat capacity is also a simple derivative

$$c_V = \left(\frac{\partial}{\partial T}\frac{3\pi^4 Nk_BT^4}{5\Theta_D^3}\right)_V = \frac{12\pi^4}{5}Nk_B\left(\frac{T}{\Theta_D}\right)^3,$$

(2.52)

which precisely explains the T^3 part of the experimental results! It turns out that the T^1 part of the experimental result for metals is explained by the electrons' contribution (see chapter 3).

Some comments are in order:

- Note from appendix 2.2.3 that Einstein's model assumes only optical phonons play a part and the optical phonons' dispersion relation is approximated by a constant.
- Note from appendix 2.2.3 that Debye's model assumes only acoustic phonons play a part and the acoustic phonons' dispersion relation is approximated by a linear relation which is essentially long-wavelength acoustic phonons.
- The above two points explain why Einstein's model is not as good as Debye's model at low temperatures. At low temperatures (and therefore low energies), only the long-wavelength acoustic phonons are excited and contribute to the heat capacity. The optical phonons are too high in energy to be excited. Thus Einstein's model is actually invalid at low temperatures.
- We also need to explain why both models can give the Dulong–Petit law at high temperatures. This is actually simple to explain: at high temperatures (which mean high energies), all phonons are excited and contribute to the heat capacity, so both models work.

2.2.3 Appendix: density of states (DOS) calculations

As the calculations for the 'quantum explanation' on section 2.2.2.2 are already quite long, this appendix consolidates the calculations of DOS for the various cases.

2.2.3.1 Born–von Karman periodic boundary conditions

First, we need to impose boundary conditions called Born–von Karman periodic boundary conditions. These periodic boundary conditions have the advantage of handling a finite solid, and yet we can avoid dealing with surface effects because now the solid is 'never ending'.

We use a 1D chain to illustrate and extrapolate the answers to 2D and 3D (figure 2.11).

Since $s = N$ and $s = 0$ are joined together, the travelling wave needs to obey[3]

$$u_N = u_0. \tag{2.53}$$

In general, this should hold:

$$u_{N+s} = u_s$$
$$| \text{ insert the solution: } u_s = ue^{iqsa - i\omega t} \tag{2.54}$$

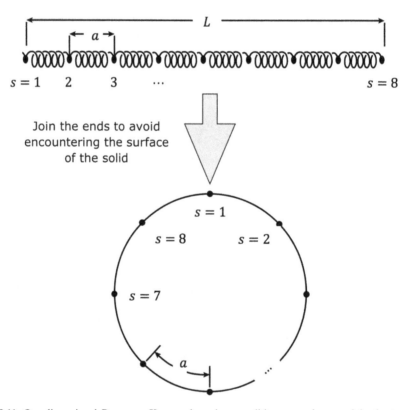

Figure 2.11. One-dimensional Born–von Karman boundary conditions are where we join the two surface atoms ($s = 0$ and $s = N$) together so that we do not encounter any surface effects.

---

[3] You can also think of the 1D chain as having the first atom labelled $s = 1$ and the last atom, $s = N$; we join them by adding a spring. Count the number of primitive cells in the joined chain in this example as having N primitive cells.

$$ue^{iq(N+s)a - i\omega t} = ue^{iqsa - i\omega t} \tag{2.55}$$

$$\Longrightarrow e^{iqNa} = 1 \tag{2.56}$$
$$\mid \text{note that } 1 = e^{\pm i2\pi n} \text{ where } n \text{ is integer}$$

$$q = 0, \ \pm\frac{2\pi}{L_x}, \ \pm\frac{4\pi}{L_x}, \ \cdots, \ \pm\frac{N\pi}{L_x} \tag{2.57}$$

These are the allowed q values (within the 1BZ) under the Born–von Karman periodic boundary conditions.[4] Note that we started with N atoms (in the joined chain), and we ended up with N allowed q values; this is a general relationship: 'the number of primitive cells N = the number of allowed q values in the 1BZ'. The 'volume' of one allowed q point is $\frac{2\pi}{L_x}$.

We shall now infer the allowed q values for 2D and 3D with N_x primitive cells in the x-direction, N_y primitive cells in the y-direction, and N_z primitive cells in the z-direction.

For 2D:
$$\begin{cases} q_x = 0, \ \pm\frac{2\pi}{L_x}, \ \pm\frac{4\pi}{L_x}, \ \cdots, \ \pm\frac{N_x\pi}{L_x} \\ q_y = 0, \ \pm\frac{2\pi}{L_y}, \ \pm\frac{4\pi}{L_y}, \ \cdots, \ \pm\frac{N_y\pi}{L_y} \end{cases}, \ \text{volume of one allowed } q \text{ point} = \frac{2\pi}{L_x}\frac{2\pi}{L_y}$$

For 3D:
$$\begin{cases} q_x = 0, \ \pm\frac{2\pi}{L_x}, \ \pm\frac{4\pi}{L_x}, \ \cdots, \ \pm\frac{N_x\pi}{L_x} \\ q_y = 0, \ \pm\frac{2\pi}{L_y}, \ \pm\frac{4\pi}{L_y}, \ \cdots, \ \pm\frac{N_y\pi}{L_y} \\ q_z = 0, \ \pm\frac{2\pi}{L_z}, \ \pm\frac{4\pi}{L_z}, \ \cdots, \ \pm\frac{N_z\pi}{L_z} \end{cases}, \ \text{volume of one allowed } q \text{ point} = \frac{2\pi}{L_x}\frac{2\pi}{L_y}\frac{2\pi}{L_z}$$

We will use Born–von Karman periodic boundary conditions again in chapter 3 for electronic wavefunctions in a solid.

Next, we explain the Debye cutoff frequency; we can then finally proceed with the DOS calculations.

2.2.3.2 The Debye cutoff frequency

- As mentioned, if there are N primitive cells (regardless of the dimension of the solid), for a monoatomic basis, there will be N modes. This means there are N allowed q points (within the 1BZ).

---

[4] Note that we always end up with odd allowed values of q for any value of N. The bottom line is: we MUST follow the rule: 'the number of primitive cells N = the number of allowed q values in the 1BZ'. So if N is odd, we count until $q = \pm\frac{(N-1)\pi}{L_x}$ and we have N allowed q values. Then, if N is even, we count until $q = \pm\frac{N\pi}{L_x}$ but both points $q = \pm\frac{N\pi}{L_x}$ are at the edge of the 1BZ and should be counted as one point. Also, note that in reality, the solid has so many atoms (primitive cells actually) that N is usually huge.

- Recall from the previous section on Born–von Karman periodic boundary conditions that the volume of one allowed q point in 3D is $\frac{(2\pi)^3}{L_xL_yL_z} = \frac{(2\pi)^3}{V}$, where $V = L_xL_yL_z$ is the volume of the solid.
- The Debye cutoff wave vector q_D is chosen to be the radius of the wave vector in reciprocal space such that the volume of the sphere $\frac{4}{3}\pi q_D^3$ is equal to the volume of N allowed q points. This is a physically motivated cutoff because physically, there are N allowed q points and there is no need to include extra q points (figure 2.12).

$$\frac{4}{3}\pi q_D^3 = N \times \frac{(2\pi)^3}{V} \tag{2.58}$$

$$q_D = \left(\frac{6\pi^2 N}{V}\right)^{1/3}$$

$$\tag{2.59}$$

| assuming theDebye model, the dispersion (for the jth branch) is $\omega = v_jq$
| so when $q = q_D$, the Debye cutoff frequency $\omega_D = v_jq_D$

$$\boxed{\omega_D = \left(\frac{6\pi^2 v_j^3 N}{V}\right)^{1/3}} \tag{2.60}$$

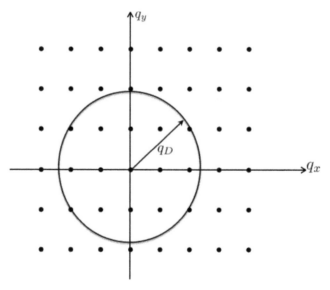

Figure 2.12. This is a 2D illustration of the Debye cutoff. Note that the dots represent allowed q points (not reciprocal space points!), so the horizontal spacing of the dots is $\frac{2\pi}{L_x}$ and the vertical spacing is $\frac{2\pi}{L_y}$. The reciprocal points are too far away to show here. The Debye cutoff wave vector q_D is chosen such that the area of the circle is equal to the area of N allowed q points, i.e. $\pi q_D^2 = N \times \frac{2\pi}{L_x}\frac{2\pi}{L_y}$.

2.2.3.3 The DOS of a 1D vibrating lattice

We start with the definition of the DOS:

$$D_j(\omega) = \sum_{q} \delta(\omega - \omega_{j\vec{q}})$$

| assuming q is dense, we convert the sum to an integral

$$(2.61)$$

$$D_j(\omega) = \frac{L_x}{2\pi} \int \delta(\omega - \omega_{j\vec{q}}) \, dq$$

| the inverse volume factor $\dfrac{L_x}{2\pi}$ compensates the volume factor dq

| change integration variable to $\omega_{j\vec{q}}$ so $dq = \dfrac{dq}{d\omega_{j\vec{q}}} d\omega_{j\vec{q}} = \dfrac{1}{v_{gj}} d\omega_{j\vec{q}}$

$$(2.62)$$

| where we recall group velocity (of branch j) is $v_{gj} = \dfrac{d\omega_{j\vec{q}}}{dq}$

$$D_j(\omega) = \frac{L_x}{2\pi} \int \delta(\omega - \omega_{j\vec{q}}) \frac{1}{v_{gj}} d\omega_{j\vec{q}}$$

| recall the definition of the delta function and carry out the integral

$$(2.63)$$

$$\boxed{D_j(\omega) = \frac{L_x}{2\pi} \frac{1}{v_{gj}} \Bigg|_{\omega_{j\vec{q}} = \omega}}$$

$$(2.64)$$

The final expression is $D_j(\omega) = \frac{L_x}{2\pi} \frac{1}{v_{gj}} \Big|_{\omega_{j\vec{q}} = \omega}$. Note that we did not specify the dispersion relation here. The group velocity v_{gj} is usually not a constant and could be some complicated function. Once the dispersion relation is known, the group velocity is calculated and, at the end, any $\omega_{j\vec{q}}$ appearing in v_{gj} is replaced by ω. As an aside, we note that in the derivation above, we assumed q is dense. Since the spacing of the allowed q points is $\frac{2\pi}{L_x}$, this implies that for q to be dense, the solid has to be large so that L_x is large.

2.2.3.4 The DOS of a 2D vibrating lattice

We are doing a very similar calculation compared to the 1D case, so we will explain less. We start with the definition of the DOS again.

$$D_j(\omega) = \sum_{q} \delta(\omega - \omega_{j\vec{q}})$$

| convert sum to integral

$$(2.65)$$

$$D_j(\omega) = \frac{L_x}{2\pi}\frac{L_y}{2\pi}\int \delta(\omega - \omega_{j\vec{q}})\, d^2\vec{q}$$

$$\Big|\ \text{use polar coordinates: } \int \cdots d^2\vec{q} = \int\int_0^{2\pi} \cdots d\theta q dq = 2\pi\int \cdots q dq$$

(2.66)

$$D_j(\omega) = 2\pi\frac{L_x}{2\pi}\frac{L_y}{2\pi}\int \delta(\omega - \omega_{j\vec{q}}) q\, dq$$

$$\Big|\ \text{change variable to } \omega_{j\vec{q}}, \text{ so } dq = \frac{1}{v_{gj}}d\omega_{j\vec{q}} \text{ and carry out the integration}$$

(2.67)

$$\boxed{D_j(\omega) = \frac{L_x L_y}{2\pi}\frac{q}{v_{gj}}\Big|_{\omega_{j\vec{q}}=\omega}}$$

(2.68)

Equation (2.68) is the final expression for the 2D DOS. Once the dispersion is known, the group velocity v_{gj} is calculated and wherever $\omega_{j\vec{q}}$ appears, it is replaced by ω.

2.2.3.5 The DOS of a 3D vibrating lattice
Again, we start with definition of the DOS.

$$D_j(\omega) = \sum_{\vec{q}}\delta(\omega - \omega_{j\vec{q}})$$

$$\Big|\ \text{convert sum to integral}$$

(2.69)

$$D_j(\omega) = \frac{L_x}{2\pi}\frac{L_y}{2\pi}\frac{L_z}{2\pi}\int \delta(\omega - \omega_{j\vec{q}})\, d^3\vec{q}$$

$$\Big|\ \text{use spherical coordinates: } \int \cdots d^3\vec{q} = \underbrace{\int_0^{2\pi} d\phi \int_0^{\pi} \sin\theta d\theta}_{=4\pi} \int \cdots q^2 dq$$

(2.70)

$$D_j(\omega) = 4\pi\frac{L_x}{2\pi}\frac{L_y}{2\pi}\frac{L_z}{2\pi}\int \delta(\omega - \omega_{j\vec{q}}) q^2\, dq$$

$$\Big|\ \text{change variable to } \omega_{j\vec{q}}, \text{ so } dq = \frac{1}{v_{gj}}d\omega_{j\vec{q}} \text{ and carry out the integration}$$

(2.71)

$$\boxed{D_j(\omega) = \frac{L_x L_y L_z}{2\pi^2}\frac{q^2}{v_{gj}}\Big|_{\omega_{j\vec{q}}=\omega}}$$

(2.72)

Equation (2.72) is the final expression for the 3D DOS. Once the dispersion is known, the group velocity v_{gj} is calculated and wherever $\omega_{j\vec{q}}$ appears, it is replaced by ω.

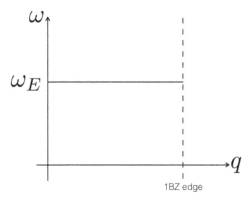

1BZ edge

Figure 2.13. Einstein's model simply assumes that the dispersion is a constant. Comparing this model with the dispersion of the diatomic chain, Einstein's model can be said to be a very crude approximation of optical modes.

2.2.3.6 The DOS of Einstein's model

This model assumes that the dispersion is simply a constant: all $\vec{q}$s have the same frequency ω_E (figure 2.13). So $\omega_{j\vec{q}} = \omega_E$. Then, the DOS,

$$D_j(\omega) = \sum_{\vec{q}} \delta(\omega - \omega_{j\vec{q}}) \tag{2.73}$$

$$D_j(\omega) = \sum_{\vec{q}} \delta(\omega - \omega_E) \tag{2.74}$$

    | for N atoms in a 3D solid, the number of allowed $\vec{q}$ in 1BZ is N

$$\boxed{D_j(\omega) = N\delta(\omega - \omega_E)} \tag{2.75}$$

2.2.3.7 The DOS of Debye's model

This model assumes that the dispersion is a linear relationship $\omega_{j\vec{q}} = v_j q$, where v_j is a constant (figure 2.14); it is also the gradient and therefore the group velocity (of the jth branch)! Now that the dispersion is specified, we simply borrow the 3D DOS expression from above.

$$D_j(\omega) = \frac{L_x L_y L_z}{2\pi^2} \frac{q^2}{v_{gj}} \bigg|_{\omega_{j\vec{q}} = \omega} \tag{2.76}$$

    | replace v_{gj} with v_j and insert $q^2 = \dfrac{\omega_{j\vec{q}}^2}{v_j^2}$

    | and let $L_x L_y L_x = V$, the volume of the solid

$$D_j(\omega) = \frac{V}{2\pi^2} \frac{\omega_{j\vec{q}}^2}{v_j^3} \bigg|_{\omega_{j\vec{q}} = \omega} \tag{2.77}$$

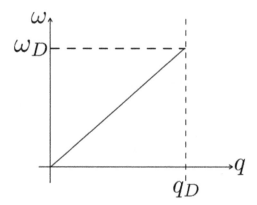

Figure 2.14. Debye's model simply assumes that the dispersion is a linear relationship. Comparing this model with the dispersion of the diatomic chain (or monoatomic chain), Debye's model can be said to be an approximation of acoustic modes in the long-wavelength limit. A physical cutoff ω_D is imposed as well.

$$D_j(\omega) = \frac{V}{2\pi^2} \frac{\omega^2}{v_j^3}$$

(2.78)

2.2.4 Some topics in lattice thermodynamics that were excluded

The following topics are not included in this book, as they require an exposition of phonon–phonon interactions, which are beyond the scope of this book.

- Thermal expansion: the harmonic approximation actually does not allow the solid to expand because the quantum expectation value of the position does not change. To model thermal expansion, we have to include anharmonic terms from the lattice potential energy.
- Thermal conductivity: the classical model of thermal conductivity simply assumes that phonons are like a gas which undergoes collisions (like small, hard particles) with other phonons, defects, and boundaries and diffuses to the cooler end of the solid.
- Establishment of thermal equilibrium: only phonon–phonon collisions can establish thermal equilibrium, as they can redistribute the energy of the phonons. There are two types of phonon–phonon collisions based on wave vector conservation: normal and Umklapp. Normal processes have their final wave vector within the 1BZ and thus do not require the final wave vector to be mapped back to the 1BZ. Umklapp processes have their final wave vector outside the 1BZ and thus require their final wave vector to be mapped back to the 1BZ. After mapping back, the final wave vector points in a somewhat altered direction and has a smaller magnitude. Umklapp processes therefore allow final wave vectors to reverse and decrease, which is important to allow equilibration to take place.

2.3 Exercises

2.3.1 Question: the spring constant

(Solution in section 6.2.1.) Taking the ionic bonding lattice energy U (with exponential model for repulsion, equation (1.47)), show that the spring constant can be written as $k = \frac{Naq^2}{4\pi\epsilon_0 R_0^2}\left(\frac{1}{\rho} - \frac{2}{R_0}\right)$. (Hint: refer to exercise 1.4.12.)

2.3.2 Question: a 1D solid, monoatomic basis, 1D vibrations

(Solution in section 6.2.2.) Now consider a 1D monoatomic vibrating chain with atoms of mass M and equal equilibrium spacing a.

1. Write down the dispersion relation (no need to derive it) and sketch it in the 1BZ, the 2BZ, and the 3BZ.
2. Explain why the minimum wavelength is $2a$ (or the maximum is $q = \frac{\pi}{a}$).
3. What about wavelengths less than $2a$? Explain using the example of wavelength $= a$.
4. Derive the group velocity expression and sketch it in the 1BZ. What happens to the group velocity at the zone boundaries?
5. Assuming the chain has only seven atoms and we impose Born–von Karman periodic boundary conditions, write down the allowed wave vectors and mark them with crosses on the dispersion relation diagram (within the 1BZ will be acceptable).
6. The speed of sound in this 1D solid is 1.08×10^4 m s^{-1}. The mass of each atom is 6.81×10^{-26} kg and the equilibrium separation a is 4.85 Å. Find the values of:
    (a) the force constant
    (b) the maximum frequency of vibration and
    (c) the Debye temperature. (Hint: you need to derive the 1D version of the Debye cutoff frequency ω_D first. Use Debye's model and set the group velocity equal to the speed of sound.)
7. If we take the 1D diatomic case (with 1D vibrations) and take both atoms (in the basis) as having the same mass M, then this would just be the 1D monoatomic case with equilibrium spacing $\frac{a}{2}$. Now reconcile the apparent differences in their dispersion relations. It is easiest to sketch the graphs and discuss the apparent differences using the graphs.

2.3.3 Question: the photon dispersion relation

(Solution in section 6.2.3.) Let us consider the dispersion relation for photons, which you actually may be familiar with.

1. Write down the dispersion relation for photons.
2. Use the formula for group velocity and check that the group velocity for photons is the value you would expect.
3. For UV light at 400 nm, what is the corresponding wave vector?[5]

---

[5] We note that the photon's wave vector is always small (as compared to the size of the BZ) and so when an electron absorbs a photon, the electron's wave vector hardly changes. This is useful for Part 3.

2.3.4 Question: 2D lattice thermodynamics

(Solution in section 6.2.4.) Now assume that we have a 2D solid. Repeat the entire calculation in section 2.2.2, starting from the quantum explanation for the Debye model and ending with the high-temperature and low-temperature limits in the 2D Debye model. Skip Einstein's model and be careful about using quantities in 2D instead of 3D.

References

[1] Kittel C 2005 *Introduction to Solid State Physics* 8th edn (New York: Wiley) https://www.wiley.com/en-us/Introduction+to+Solid+State+Physics%2C+8th+Edition-p-9780471415268

[2] Ibach H and Lüth H 2009 *Solid-State Physics: An Introduction to Principles of Materials Science* (Berlin: Springer)

Chapter 3

Electronic properties

This chapter introduces the basic electronic properties of solids. 'Electronic proper-ties' refers to the energy levels (energy eigenvalues) of the electrons, the thermody-namical properties of the electrons, and the electrical conductivity of the solid.

In our discussion of the energy levels of the electrons, we start with the simplest quantum model, the Sommerfeld model; then, in section 3.2, we cover various improved quantum models. The most important physical consequence is the appearance of an energy bandgap.

In our discussion of the thermodynamical properties, we cover heat capacity and thermal conductivity. Heat capacity is discussed using the classical Drude model and the quantum Sommerfeld model. We then explicitly show that the quantum model explains the low-temperature behaviour of the heat capacity.

We discuss the electrical conductivity using the classical Drude model but do not discuss it further, as the quantum model is beyond the scope of this book.

3.1 Free electron gas

3.1.1 Classical treatment: the Drude model

This section covers the Drude model of metals. The Drude model was the first model for solids (specifically metals) in which the 'sea' of electrons was treated as a gas of classical (negatively charged) particles.

This gas moves among (positively charged) ions. The electrons collide instanta-neously with these ions and their momenta change. The electrons are assumed not to interact with each other.

Although the Drude model is now considered a failure, it had some success with the Wiedemann–Franz ratio, and, more importantly, it paved the way for quantum models.

We start with the first physical property to be calculated using the Drude model, namely electrical conductivity.

3.1.1.1 Transport properties: electrical conductivity: Ohm's law

The microscopic expression for the electronic current density $\vec{J}$ (current per cross sectional area) is:

$$\vec{J} = n(-e)\vec{v} = -\frac{ne}{m}\vec{p} \tag{3.1}$$

where n is the electron density; m is the electron's mass; e is a positive number, namely 1.602×10^{-19} C; $\vec{v}$ is the average (or drift) velocity; and $\vec{p}$ is the average momentum.

The application of the electric field results in a force $\vec{F} = -e\vec{E}$ that acts on the electron. Consider a short time δt and define $\frac{1}{\tau}$ as the rate (or frequency) of collision.

- No collision in $\delta t \Rightarrow$ probability is $\left(1 - \frac{\delta t}{\tau}\right) \Rightarrow$ momentum increase $\delta \vec{p} = \vec{F}\delta t$
- Collision in $\delta t \Rightarrow$ probability is $\frac{\delta t}{\tau} \Rightarrow$ momentum is lost (or randomised) then at most $\vec{F}\delta t$ is regained

The contribution to the average momentum (to the first order in δt) is:

$$\vec{p}(t + \delta t) = \left(1 - \frac{\delta t}{\tau}\right)(\vec{p} + \vec{F}\delta t) + \frac{\delta t}{\tau}\left(\underbrace{0}_{\text{randomise to zero}} + \vec{F}\delta t\right)$$

| first term: 'did not collide', $\qquad$ (3.2)
| second term: 'collided'
| expand and keep to first order in δt

$$\vec{p}(t + \delta t) - \vec{p}(t) = -\frac{\delta t}{\tau}\vec{p}(t) + \vec{F}\delta t$$

$\qquad$ (3.3)

| note that effectively, electrons that didn't collide contributed

$$\lim_{\delta t \to 0} \frac{\vec{p}(t + \delta t) - \vec{p}(t)}{\delta t} = -\frac{\vec{p}(t)}{\tau} + \vec{F} \tag{3.4}$$

$$\frac{d\vec{p}(t)}{dt} = -\frac{\vec{p}(t)}{\tau} + \vec{F}$$

| at steady state, $\dfrac{d\vec{p}(t)}{dt} = 0$, $\qquad$ (3.5)

| which means the average momentum does not change

$$0 = -\frac{\vec{p}(t)_{ss}}{\tau} - e\vec{E} \tag{3.6}$$

$$\vec{p}(t)_{ss} = -e\tau\vec{E} \tag{3.7}$$

$$\Longrightarrow \vec{J} = -\frac{ne}{m}\vec{p}(t)_{ss} \tag{3.8}$$

$$|\text{then } \vec{p}(t)_{ss} = e\vec{E}\tau \tag{3.9}$$

$$\vec{J} = \frac{ne^2\tau}{m}\vec{E} \tag{3.10}$$

| electrical conductivity is defined by Ohm's law: $\vec{J} = \sigma\vec{E}$

$$\Longrightarrow \sigma = \frac{ne^2\tau}{m} \tag{3.11}$$

Conductivity in the Drude model: $\quad \sigma = \dfrac{ne^2\tau}{m} \tag{3.12}$

Let us do a quick order of magnitude check:
- a typical metal's conductivity $\sigma \approx 1 \times 10^{-8} \ \Omega^{-1} \ \text{m}^{-1}$
- a typical metal's carrier density $n \approx 10^{22} \ \text{cm}^{-3}$
- thus $\tau \approx 5.68 \times 10^{-13}$ s
- since the electron gas is treated as a classical gas, we have $\frac{1}{2}m\langle v^2\rangle = \frac{3}{2}k_B T$, so for $T \approx 300$ K, we have $\sqrt{\langle v^2\rangle} \approx 1.167 \times 10^5 \ \text{m s}^{-1}$
- the mean free path is $\sqrt{\langle v^2\rangle}\tau \approx 6.63 \times 10^{-8}$ m, which is of the right order for interatomic distances.

So the Drude model is somewhat successful here. It turns out that the electronic velocity $\sqrt{\langle v^2\rangle} \approx 1.167 \times 10^5 \ \text{m s}^{-1}$ is a severe underestimate.

Next, we talk about the thermodynamics of electrons using the Drude model.

3.1.1.2 Thermal properties: heat capacity, thermal conductivity, and the Wiedemann–Franz ratio

Heat capacity: Since the electron gas is treated as classical, the standard Dulong–Petit-like result $c_V = \frac{3}{2}nk_B$, which is independent of temperature, applies.[1] The low-temperature heat capacities of metals show a temperature-dependent relationship (section 2.2.1)

$$c_V = \gamma T + \alpha T^3 \tag{3.13}$$

---

[1] See the derivation in statistical mechanics books. Recall also that in section 2.2.2, the lattice vibrations were treated classically. That is effectively a classical non-interacting phonon gas problem, which explains why the electronic heat capacity and the lattice heat capacity are so similar.

and the term αT^3 was already explained in section 2.2.2 as the contribution from the lattice. The term γT must be the (low-temperature) electronic contribution, but the Drude model has no way to explain it.

Thermal conductivity: the thermal conductivity κ is defined by Fourier's law:

$$\vec{J}_{\text{heat}} = -\kappa \vec{\nabla} T, \tag{3.14}$$

where $\vec{J}_{\text{heat}}$ is the heat flux density (energy per second per unit area) and $\vec{\nabla} T$ is the temperature gradient. As this is treated as a classical gas, the classical kinetic theory result $\kappa = \frac{1}{3} c_V \sqrt{\langle v^2 \rangle} \, l$ applies. The heat capacity is $c_V = \frac{3}{2} n k_B$, the average speed is the root-mean-square $\sqrt{\langle v^2 \rangle}$ calculated from $\frac{1}{2} m \langle v^2 \rangle = \frac{3}{2} k_B T$, and the mean free path is $l = \sqrt{\langle v^2 \rangle} \, \tau$.

The Wiedemann–Franz ratio: The Wiedemann–Franz ratio is defined as $\frac{\kappa}{\sigma}$, so

$$\frac{\kappa}{\sigma} = \frac{\frac{1}{3} c_V \sqrt{\langle v^2 \rangle} \, l}{\frac{ne^2 \tau}{m}} = \frac{\frac{1}{3} \frac{3}{2} n k_B \langle v^2 \rangle \tau}{\frac{ne^2 \tau}{m}} = \frac{\frac{1}{2} n k_B \frac{3 k_B T}{m} \tau}{\frac{ne^2 \tau}{m}} = \frac{3}{2} \frac{k_B^2}{e^2} T \tag{3.15}$$

$$\implies \frac{\kappa}{\sigma T} = \frac{3}{2} \frac{k_B^2}{e^2} = 1.1 \times 10^{-8} \text{ W } \Omega \text{ K}^{-2} \tag{3.16}$$

The experimental value for $\frac{\kappa}{\sigma T}$ for many metals $\approx 2.5 \times 10^{-8}$ W Ω K^{-2} so this calculation gave a coincidental agreement! We cannot really trust this agreement, as the Drude model cannot explain anything else.

Finally, we look at the classical Hall effect in the context of the Drude model.

3.1.1.3 Transport properties: the classical Hall effect
The Hall effect is the effect that occurs when a magnetic field is applied to a metal which is connected to a circuit (this means the metal is in an electric field). Recall equation (3.5):

$$\frac{d\vec{p}}{dt} = -\frac{\vec{p}(t)}{\tau} + \vec{F}(t) \tag{3.17}$$

$\qquad$ | the force now is the Lorentz force $\vec{F} = -e\vec{E} - e\vec{v} \times \vec{B}$

$$m\left(\frac{d\vec{v}}{dt} + \frac{\vec{v}}{\tau}\right) = -e\vec{E} - e\vec{v} \times \vec{B}$$

$\qquad$ | assume a DC magnetic field $\vec{B} = B\hat{z}$

$\qquad$ | assume a temporally varying field $\vec{E} = \begin{pmatrix} E_x \\ E_y \\ E_z \end{pmatrix} e^{i\omega t}$ $\qquad$ (3.18)

$\qquad$ | thesolution ansatz is a similar form, $\vec{v} = \begin{pmatrix} v_x \\ v_y \\ v_z \end{pmatrix} e^{i\omega t}$

$$m\frac{d}{dt}\begin{pmatrix} v_x e^{i\omega t} \\ v_y e^{i\omega t} \\ v_z e^{i\omega t} \end{pmatrix} + \frac{m}{\tau}\begin{pmatrix} v_x e^{i\omega t} \\ v_y e^{i\omega t} \\ v_z e^{i\omega t} \end{pmatrix} = -e\begin{pmatrix} E_x e^{i\omega t} \\ E_y e^{i\omega t} \\ E_z e^{i\omega t} \end{pmatrix} - e\begin{pmatrix} v_x e^{i\omega t} \\ v_y e^{i\omega t} \\ v_z e^{i\omega t} \end{pmatrix} \times \begin{pmatrix} 0 \\ 0 \\ B \end{pmatrix} \qquad (3.19)$$

$$i\omega m\begin{pmatrix} v_x \\ v_y \\ v_z \end{pmatrix} + \frac{m}{\tau}\begin{pmatrix} v_x \\ v_y \\ v_z \end{pmatrix} = -e\begin{pmatrix} E_x \\ E_y \\ E_z \end{pmatrix} - e\begin{pmatrix} Bv_y \\ -Bv_x \\ 0 \end{pmatrix} \qquad (3.20)$$

$$(1 + i\omega\tau)\begin{pmatrix} v_x \\ v_y \\ v_z \end{pmatrix} = -\frac{e\tau}{m}\begin{pmatrix} E_x + Bv_y \\ E_y - Bv_x \\ E_z \end{pmatrix} \qquad (3.21)$$

We now make a simplifying assumption just to reduce the algebra. We assume the electric field is produced by DC, so we set $\omega = 0$

$$\begin{pmatrix} v_x \\ v_y \\ v_z \end{pmatrix} = -\frac{e\tau}{m}\begin{pmatrix} E_x \\ E_y \\ E_z \end{pmatrix} - \frac{e\tau B}{m}\begin{pmatrix} v_y \\ -v_x \\ 0 \end{pmatrix} \qquad (3.22)$$

| define the cyclotron frequency $\omega_c = \dfrac{eB}{m}$

| and solve the simultaneous equations

$$\begin{cases} v_x = -\dfrac{e\tau}{m}\dfrac{(E_x - \omega_c\tau E_y)}{1 + \omega_c^2\tau^2} \\[2mm] v_y = -\dfrac{e\tau}{m}\dfrac{(\omega_c\tau E_x + E_y)}{1 + \omega_c^2\tau^2} \\[2mm] v_z = -\dfrac{e\tau}{m}E_z \end{cases} \qquad (3.23)$$

So we are invited to define a magnetoconductivity matrix $\overleftrightarrow{\sigma}$ and a generalised Ohm's law:

$$\vec{J} = \overleftrightarrow{\sigma}\vec{E} \qquad (3.24)$$

$$\implies -ne\vec{v} = \overleftrightarrow{\sigma}\vec{E} \qquad (3.25)$$

$$\begin{pmatrix} v_x \\ v_y \\ v_z \end{pmatrix} = -\frac{1}{ne}\begin{pmatrix} \sigma_{xx} & \sigma_{xy} & \sigma_{xz} \\ \sigma_{yx} & \sigma_{yy} & \sigma_{yz} \\ \sigma_{zx} & \sigma_{zy} & \sigma_{zz} \end{pmatrix}\begin{pmatrix} E_x \\ E_y \\ E_z \end{pmatrix} \qquad (3.26)$$

Comparing this with the earlier expressions for v_x, v_y, and v_z, we have (exercise)

$$\overleftrightarrow{\sigma} = \begin{pmatrix} \dfrac{\sigma_D}{1+\omega_c^2\tau^2} & -\dfrac{\sigma_D\omega_c\tau}{1+\omega_c^2\tau^2} & 0 \\[3mm] \dfrac{\sigma_D\omega_c\tau}{1+\omega_c^2\tau^2} & \dfrac{\sigma_D}{1+\omega_c^2\tau^2} & 0 \\[3mm] 0 & 0 & \sigma_D \end{pmatrix},$$

(3.27)

where $\sigma_D = \frac{ne^2\tau}{m}$ is the Drude conductivity.

To discuss the Hall effect, we pick the specific setup shown in figure 3.1 and, in the DC or $\omega = 0$ case, charges accumulate on the surfaces normal to the y-direction until an E_y field exactly cancels the Lorentz force; then, $J_y = 0$, so

$$\begin{pmatrix} J_x \\ J_y \\ J_z \end{pmatrix} = \frac{\sigma_D}{1+\omega_c^2\tau^2} \begin{pmatrix} 1 & -\omega_c\tau & 0 \\ \omega_c\tau & 1 & 0 \\ 0 & 0 & 1+\omega_c^2\tau^2 \end{pmatrix} \begin{pmatrix} E_x \\ E_y \\ E_z \end{pmatrix}$$

(3.28)

$$\Longrightarrow J_y = \frac{\sigma_D}{1+\omega_c^2\tau^2}(E_x\omega_c\tau + E_y) = 0$$

(3.29)

$$\Longrightarrow E_y = -E_x\omega_c\tau$$

(3.30)

Hall coefficient R is defined as $R = \dfrac{E_y}{J_xB}$

(3.31)

$$R = \frac{-E_x\omega_c\tau}{\dfrac{\sigma_D}{1+\omega_c^2\tau^2}(E_x - E_y\omega_c\tau)B}$$

(3.32)

$$\mid \text{use } E_y = -E_x\omega_c\tau, \ \omega_c = \frac{eB}{m} \text{ and } \sigma_D = \frac{ne^2\tau}{m}$$

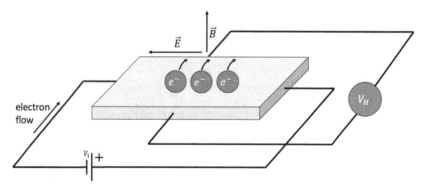

Figure 3.1. Setup used for the discussion of the (classical) Hall effect.

$$R = -\frac{\omega_c \tau}{\sigma_D B} \tag{3.33}$$

$$R = -\frac{e\tau}{m} \times \frac{m}{ne^2\tau} \tag{3.34}$$

$$R = \frac{1}{n(-e)} \tag{3.35}$$

Thus, it is useful to measure the carrier concentration using the Hall effect. However, experiments have found that R actually varies with the magnetic field, and in many metals, R is actually positive. We will briefly explain how R can be positive using the so-called band theory later.

It should be noted that the recent phenomena of the integer quantum Hall effect and the fractional quantum Hall effect won the Nobel prize and are active areas of research today, so there is still a lot more to say about the Hall effect.

3.1.2 Quantum treatment: the Sommerfeld model

In principle, the quantum treatment of solids involves stating the Hamiltonian operator for the entire solid. It is actually quite easy to state the Hamiltonian operator, which is $H = H_{solid} + H_{external}$, where:

- H_{solid} is the Hamiltonian operator for the solid. This consists of five separate terms, i.e. $H_{solid} = T_{I-I} + T_{el} + V_{I-I} + V_{el-el} + V_{el-I}$:[2]
  - T_{I-I} is the kinetic energy operator for the ions.[2]
  - T_{el} is the kinetic energy operator for the electrons.[3]
  - V_{I-I} is the inter-ionic potential energy operator. This could be coulombic in form.
  - V_{el-el} is the inter-electronic potential energy operator. This could be coulombic in form.
  - V_{el-I} is the ion–electron potential energy operator. This could be coulombic in form.
- $H_{external}$ is the Hamiltonian operator representing the interaction between external fields and the solid. This term can be time dependent.

Then, again in principle, the whole of solid-state physics is obtained when the time-dependent Schrödinger equation is solved:

$$i\hbar\frac{\partial}{\partial t}\psi(\vec{r}, t) = H\psi(\vec{r}, t). \tag{3.36}$$

It turns out that it is unsolvable and so we need to make approximations. The simplest model is called the Sommerfeld model, in which we neglect T_{I-I} (frozen

---

[2] We define an ion as the combination of the nucleus and the core electrons. This is the so-called the rigid ion model.

[3] Here, 'electrons' refers to valence electrons.

ions), we neglect V_{I-I} (there is no inter-ionic interaction), we neglect V_{el-el} (there is no inter-electronic interaction) and we neglect V_{el-I} (effectively, there are no ions).

This is obviously very artificial, as it assumes that the electrons in the metal are behaving like a gas of non-interacting (free) fermions. The background lattice and the coulombic interactions between the electrons are not really small or negligible.

Nevertheless, we will push on with the Sommerfeld model to see whether it can explain the electronic heat capacity of a metal. It turns out to be quite acceptable![4]

3.1.2.1 Warm up: a free electron Fermi gas in a 1D box

As a warm-up to the 3D free electron gas (Sommerfeld model), let us first work out the 1D free electron gas. Let the non-interacting electron gas be trapped in a box of length L_x. This is the 1D infinite square well problem in quantum mechanics (figure 3.2).

As this is a stationary system, we use the time-independent Schrödinger equation

$$-\frac{\hbar^2}{2m}\frac{d^2\psi(x)}{dx^2} = E\psi(x) \tag{3.37}$$

$$\frac{d^2\psi(x)}{dx^2} = -k^2\psi(x) \quad \text{where } k^2 = \frac{2mE}{\hbar^2}. \tag{3.38}$$

$$\text{Solving gives} \quad \psi(x) = Ae^{ikx} + Be^{-ikx}. \tag{3.39}$$

Since the wave function must vanish at the two walls of the well, the boundary conditions are $\psi(0) = 0$ and $\psi(L_x) = 0$. Substituting into $\psi(x)$,

$$\Rightarrow A + B = 0 \quad \text{and} \quad \Rightarrow Ae^{ikL_x} + Be^{-ikL_x} = 0 \tag{3.40}$$

$$\text{together gives,} \quad A(e^{ikL_x} - e^{-ikL_x}) = 0 \tag{3.41}$$

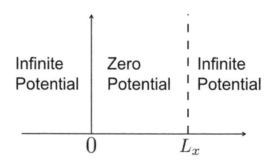

Figure 3.2. The 1D infinite well of width L_x.

---

[4] The underlying reasons for such a simple model that actually has some use are quite deep and are not discussed here.

$$\sin(kL_x) = 0 \tag{3.42}$$

$kL_x = l\pi$ where l is an integer, $l = 1, 2, 3...$
| note that there is no $l = 0$ state in the square well
$$\tag{3.43}$$

putting $k = \dfrac{l\pi}{L_x}$ into E, $\Rightarrow E = \dfrac{\hbar^2}{2m}\left(\dfrac{l\pi}{L_x}\right)^2$ $\tag{3.44}$

after normalization, $\psi(x) = \sqrt{\dfrac{2}{L_x}}\sin(kx) = \sqrt{\dfrac{2}{L_x}}\sin\left(\dfrac{l\pi}{L_x}x\right).$ $\tag{3.45}$

In the case of a solid, as will be explained on section 3.2.1.4, it is advantageous to use the Born–von Karman periodic boundary condition:[5] $\psi(x) = \psi(x + L_x)$. Thus,

$\Rightarrow \sin(kx) = \sin(k(x + L_x))$
| use the trigonometric addition formula
$$\tag{3.46}$$

$$\sin(kx) = \sin(kx)\cos(kL_x) + \cos(kx)\sin(kL_x) \tag{3.47}$$

this means that $\cos(kL_x) = 1$, $\sin(kL_x) = 0$, $\tag{3.48}$

which means that $kL_x = (\text{even integer})\pi = 2n\pi$,   where $n = 1, 2, ...$ $\tag{3.49}$

$$k = \frac{2\pi}{L_x}, \frac{4\pi}{L_x}, ... \tag{3.50}$$

- Under the Pauli exclusion principle, each level can hold two electrons. We say each level has a degeneracy of two.
- If we have N electrons and the topmost filled level is called n_F, they are related as follows: $2n_F = N$.
- The Fermi energy ε_F is the energy of the topmost filled level:

$$\varepsilon_F = \frac{\hbar^2}{2m}k_F^2 = \frac{\hbar^2}{2m}\left(\frac{2n_F\pi}{L_x}\right)^2 = \frac{\hbar^2}{2m}\left(\frac{N\pi}{L_x}\right)^2 \tag{3.51}$$

- This whole discussion is at temperature $T = 0$ actually. This is called the ground state. We shall now discuss the effects of temperature in order to work out the electronic heat capacity.

---

[5] This condition is used to maintain the translational symmetry of the solid (with a finite number of sites) and avoid any need to discuss surface effects.

3.1.2.2 The effect of temperature: the Fermi–Dirac distribution

The Fermi–Dirac distribution gives the occupation probability of a state at energy ε at thermal equilibrium for non-interacting fermions[6] (figure 3.3).

$$f(\varepsilon, T) = \frac{1}{e^{(\varepsilon-\mu)/k_B T} + 1},$$ (3.52)

where k_B is the Boltzmann constant.

- The quantity μ is called the chemical potential and is a function of temperature.
- The definition used for its derivation is that μ is such that the total number of particles remains constant at N at temperature T. As will be derived later (equation (3.83)), in the region where $\mu \gg k_B T$, the expression for μ is as follows:

$$\mu \approx \varepsilon_F \left[1 - \frac{\pi^2}{12} \left(\frac{k_B T}{\mu} \right)^2 \right]$$ (3.53)

- At $T = 0$, $\mu = \varepsilon_F$. At other temperatures, $\mu = \varepsilon$, where $f(\varepsilon, T) = \frac{1}{2}$.

3.1.2.3 A 3D free electron Fermi gas (Sommerfeld model)

Assume now that the electrons in a metal are treated as an electron gas confined in a rectangular box whose dimensions are L_x by L_y by L_z. The 3D (time-independent) Schrödinger equation is

$$-\frac{\hbar^2}{2m} \vec{\nabla}^2 \psi(\vec{r}) = E\psi(\vec{r}).$$ (3.54)

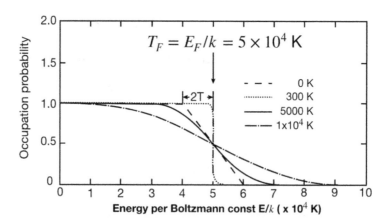

Figure 3.3. The Fermi–Dirac distributions for various temperatures.

---

[6] As this distribution is derived in the 'Statistical Mechanics' course, we shall not derive it here.

Working in Cartesian coordinates and using the separable form $\psi(\vec{r}) = \psi(x)\psi(y)\psi(z)$, the solution (from any quantum mechanics textbook) is

$$\psi(\vec{r}) = \sqrt{\frac{2}{L_x}} \sin(k_x x)\sqrt{\frac{2}{L_y}} \sin(k_y y)\sqrt{\frac{2}{L_z}} \sin(k_z z) \tag{3.55}$$

$$E(\vec{k}) = \frac{\hbar^2}{2m}k_x^2 + \frac{\hbar^2}{2m}k_y^2 + \frac{\hbar^2}{2m}k_z^2 = \frac{\hbar^2}{2m}\vec{k}^2 \tag{3.56}$$

with $\quad k_x = 0, \pm\dfrac{2\pi}{L_x}, \pm\dfrac{4\pi}{L_x}..., \quad k_y = 0, \pm\dfrac{2\pi}{L_y}, \pm\dfrac{4\pi}{L_y}..., \quad k_z = 0, \pm\dfrac{2\pi}{L_z}, \pm\dfrac{4\pi}{L_z}... \tag{3.57}$

[7]where the Born–von Karman periodic boundary conditions $\psi(x + L_x) = \psi(x)$, $\psi(y + L_y) = \psi(y)$, and $\psi(z + L_z) = \psi(z)$ are implemented (figure 3.4).
- Each k-point holds two electrons due to the Pauli exclusion principle. If we have N electrons, they fill k-space in a Fermi sphere of radius k_F.
- The Fermi energy is the energy associated with k_F: $\varepsilon_F = \dfrac{\hbar^2}{2m}\vec{k}_F^2$.
- Each k-point has a volume of $\dfrac{2\pi}{L_x}\dfrac{2\pi}{L_y}\dfrac{2\pi}{L_z}$, so the total number of electrons N is

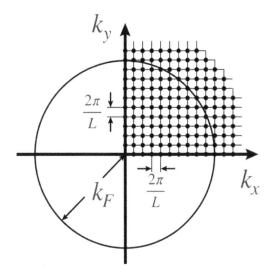

Figure 3.4. A 2D Fermi sphere with a quadrant of points shown. These points are allowed under the Born–von Karman periodic boundary conditions.

---

[7] If you compare this with the 3D quantum box, that problem only assumes positive k. The condensed matter's k is for reciprocal space, which has both positive and negative values.

$$N = 2 \times \frac{\frac{4}{3}\pi k_F^3}{\frac{2\pi}{L_x}\frac{2\pi}{L_y}\frac{2\pi}{L_z}}$$

(3.58)

| the factor of two appearsbecause each k − point can hold two electrons.

| Let the volume of the solid be $V = L_x L_y L_z$

$$k_F = \left(\frac{3\pi^2 N}{V}\right)^{1/3}$$

(3.59)

| let $n = \dfrac{N}{V}$ be the electron density;

$$k_F = (3\pi^2 n)^{1/3}.$$

(3.60)

So the Fermi wave vector k_F only depends on the density of electrons.

- Then the Fermi energy $\varepsilon_F = \frac{\hbar^2}{2m}\vec{k}_F^2 = \frac{\hbar^2}{2m}(3\pi^2 n)^{2/3}$.

- The Fermi velocity v_F can be defined as $v_F = \frac{\hbar k_F}{m} = \frac{\hbar}{m}(3\pi^2 n)^{1/3}$.

- The Fermi temperature T_F is defined as $T_F = \frac{\varepsilon_F}{k_B} = \frac{\hbar^2}{2mk_B}(3\pi^2 n)^{2/3}$. Note that this has nothing to do with the temperature of the gas. The temperature of the gas at this point in the discussion is actually zero. The idea of the Fermi temperature is to set an order of magnitude: for a Fermi gas at temperature T, the fraction of the electrons can be excited thermally is $\frac{T}{T_F}$.

- The density of states (DOS) $D(\varepsilon)$ is the number of states with energy ε:

$$D(\varepsilon) = 2\sum_{\vec{k}}\delta(\varepsilon - E(\vec{k}))$$

| the factor of two is due to spin degeneracy.

(3.61)

| This expression counts all states forwhich $\varepsilon = E(\vec{k})$.

| Assuming k is dense, we can change the summation to integral.

$$D(\varepsilon) = 2\frac{L_x}{2\pi}\frac{L_y}{2\pi}\frac{L_z}{2\pi}\int\delta(\varepsilon - E(\vec{k}))\ d^3\vec{k}$$

(3.62)

| go to spherical coordinates and integrate over θ and ϕ to get 4π

$$D(\varepsilon) = 2\frac{V}{(2\pi)^3}\int_0^\infty \delta\left(\varepsilon - \frac{\hbar^2 k^2}{2m}\right)k^2 4\pi dk$$

(3.63)

| change variable, let $\varepsilon' = \dfrac{\hbar^2 k^2}{2m}$ so $dk = \dfrac{dk}{d\varepsilon'}d\varepsilon' = \sqrt{\dfrac{2m}{\hbar^2}}\dfrac{1}{2\sqrt{\varepsilon'}}d\varepsilon'$

$$D(\varepsilon) = \frac{V}{(2\pi)^3} 4\pi \left(\frac{2m}{\hbar^2}\right)^{3/2} \int_0^\infty \delta(\varepsilon - \varepsilon') \sqrt{\varepsilon'}\, d\varepsilon' \tag{3.64}$$

$$D(\varepsilon) = \frac{V}{2\pi^2} \left(\frac{2m}{\hbar^2}\right)^{3/2} \sqrt{\varepsilon}. \tag{3.65}$$

k is dense in the sense that the rectangular box is large, so L_x, L_y and L_z are large. Also, the inverse (k-space) volume factor $\frac{L_x}{2\pi}\frac{L_y}{2\pi}\frac{L_z}{2\pi}$ compensates the (k-space) volume factor $d^3\vec{k}$.

- Note that if we have $D(\varepsilon)$, then the total number of electrons N (at $T = 0$) can be calculated by

$$N = \int_0^{\varepsilon_F} D(\varepsilon) d\varepsilon \tag{3.66}$$

$$N = \int_0^{\varepsilon_F} \frac{V}{2\pi^2} \left(\frac{2m}{\hbar^2}\right)^{3/2} \sqrt{\varepsilon}\, d\varepsilon \tag{3.67}$$

$$N = \frac{V}{2\pi^2} \left(\frac{2m}{\hbar^2}\right)^{3/2} \frac{2}{3} [\varepsilon^{3/2}]_0^{\varepsilon_F} \tag{3.68}$$

$$\varepsilon_F = \frac{\hbar^2}{2m} \left(\frac{3\pi^2 N}{V}\right)^{2/3}, \tag{3.69}$$

which is of course consistent with the earlier result.

- The incorporation of temperature allows us to use the Fermi–Dirac distribution shown in section 3.1.2.2. Now that we know how to include temperature in the discussion of free electrons (in the quantum theory), we will next discuss the thermodynamical properties of the electron gas.

3.1.2.4 The specific heat capacity of the 3D free electron Fermi gas

[8]Historically, the low-temperature specific heat capacity of metals deviates significantly from the equipartition theorem, which is classical, as seen in section 3.1.1.2. This implies that the electron gas needs to be treated quantum mechanically in order to explain this behaviour.

The low-temperature experimental specific heat follows

$$c_V = \gamma T + \alpha T^3, \tag{3.70}$$

---

[8] This section is covered in greater detail in many statistical mechanics textbooks.

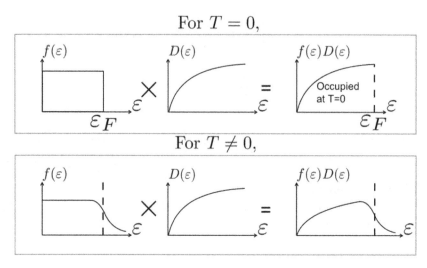

Figure 3.5. The effect of temperature on the quantum free electron gas.

where γ and α are constants. The cubic term is the phonon contribution that you learnt about in section 2.2.2. This section aims to understand the linear term (figure 3.5).

The internal energy $U(0)$ at zero temperature is

$$U(0) = \int_0^\infty \varepsilon D(\varepsilon) f(\varepsilon, T = 0) d\varepsilon \qquad (3.71)$$

| note that $f(\varepsilon, T = 0) = \theta(\varepsilon - \varepsilon_F)$, the step function (see figure 3.5)

$$U(0) = \frac{V}{2\pi^2}\left(\frac{2m}{\hbar^2}\right)^{3/2} \int_0^{\varepsilon_F} \varepsilon^{3/2} d\varepsilon \qquad (3.72)$$

$$U(0) = \frac{V}{5\pi^2}\left(\frac{2m}{\hbar^2}\right)^{3/2} \varepsilon_F^{5/2}. \qquad (3.73)$$

The internal energy $U(T)$ at temperature T is

$$U(T) = \int_0^\infty \varepsilon D(\varepsilon) f(\varepsilon, T) d\varepsilon \qquad (3.74)$$

$$U(T) = \frac{V}{2\pi^2}\left(\frac{2m}{\hbar^2}\right)^{3/2} \int_0^\infty \frac{\varepsilon^{3/2}}{e^{(\varepsilon - \mu)/k_B T} + 1} d\varepsilon \qquad (3.75)$$

| this integral is known as a Fermi–Dirac integral (such integrals have no exact forms).
| Use the asymptotic form for the integral.

$$U(T) \approx \frac{V}{2\pi^2}\left(\frac{2m}{\hbar^2}\right)^{3/2}\frac{2}{5}\mu^{5/2}\left(1 + \frac{\pi^2\frac{3}{2}\left(\frac{3}{2}+1\right)}{6\frac{\mu^2}{(k_BT)^2}}\right) \tag{3.76}$$

$$U(T) = \frac{V}{2\pi^2}\left(\frac{2m}{\hbar^2}\right)^{3/2}\frac{2}{5}\mu^{5/2}\left(1 + \frac{5}{8}\pi^2\frac{(k_BT)^2}{\mu^2}\right) \tag{3.77}$$

[9]We need the temperature dependence of μ. By definition, μ is determined by the constraint that the total number of electrons is constant (at temperature T).

$$N = \int_0^\infty D(\varepsilon)f(\varepsilon, T)d\varepsilon \tag{3.79}$$

$$N = \frac{V}{2\pi^2}\left(\frac{2m}{\hbar^2}\right)^{3/2}\int_0^\infty \frac{\varepsilon^{1/2}}{e^{(\varepsilon-\mu)/k_BT} + 1}d\varepsilon \tag{3.80}$$

| which is the $j = \dfrac{1}{2}$ Fermi–Dirac integral, where $y_0 = \dfrac{\mu}{k_BT}$

$$N \approx \frac{V}{2\pi^2}\left(\frac{2m}{\hbar^2}\right)^{3/2}\frac{2}{3}\mu^{3/2}\left(1 + \frac{\pi^2\frac{1}{2}\left(\frac{1}{2}+1\right)}{6\frac{\mu^2}{(k_BT)^2}}\right) \tag{3.81}$$

| recall that $\varepsilon_F = \dfrac{\hbar^2}{2m}\left(\dfrac{3\pi^2N}{V}\right)^{2/3}$

$$\varepsilon_F = \mu\left(1 + \frac{\pi^2}{8}\left(\frac{k_BT}{\mu}\right)^2\right)^{2/3} \tag{3.82}$$

$$\mu \approx \varepsilon_F\left(1 - \frac{2}{3}\frac{\pi^2}{8}\left(\frac{k_BT}{\mu}\right)^2\right) \tag{3.83}$$

---

[9] The asymptotic form in the limit $\mu \approx \varepsilon_F \gg k_BT$ is given as

$$F_j(y_0) = \int_0^\infty \frac{y^j}{e^{(y-y_0)} + 1}dy \approx \frac{y_0^{j+1}}{j+1}\left(1 + \frac{\pi^2 j(j+1)}{6y_0^2} + \cdots\right) \tag{3.78}$$

In this case, we have $y_0 = \frac{\mu}{k_BT}$ and $j = \frac{3}{2}$.

$$\mu^{5/2} \approx \varepsilon_F^{5/2}\left(1 - \frac{5}{2}\frac{\pi^2}{12}\left(\frac{k_B T}{\mu}\right)^2\right) \tag{3.84}$$

Going back to $U(T)$ and keeping only the terms up to $\left(\frac{k_B T}{\varepsilon_F}\right)^2$,

$$U(T) \approx \frac{V}{2\pi^2}\left(\frac{2m}{\hbar^2}\right)^{3/2}\frac{2}{5}\varepsilon_F^{5/2}\left(1 - \frac{5\pi^2}{24}\left(\frac{k_B T}{\mu}\right)^2\right)\left(1 + \frac{5}{8}\pi^2\frac{(k_B T)^2}{\mu^2}\right) \tag{3.85}$$

$$U(T) \approx \frac{V}{2\pi^2}\left(\frac{2m}{\hbar^2}\right)^{3/2}\frac{2}{5}\varepsilon_F^{5/2}\left(1 - \frac{5\pi^2}{24}\left(\frac{k_B T}{\mu}\right)^2 + \frac{5\pi^2}{8}\frac{(k_B T)^2}{\mu^2}\right) \tag{3.86}$$

$$U(T) = \frac{V}{2\pi^2}\left(\frac{2m}{\hbar^2}\right)^{3/2}\frac{2}{5}\varepsilon_F^{5/2}\left(1 + \frac{5\pi^2}{12}\left(\frac{k_B T}{\mu}\right)^2\right) \tag{3.87}$$

| we make the last approximation that $\mu \approx \varepsilon_F$;
| note that the first term is $U(0)$

$$U(T) - U(0) \approx \frac{V}{12}\left(\frac{2m}{\hbar^2}\right)^{3/2}\sqrt{\varepsilon_F}k_B^2 T^2 \tag{3.88}$$

| replace $\left(\frac{2m}{\hbar^2}\right)^{3/2} = \varepsilon_F^{-3/2}\frac{3\pi^2 N}{V}$

$$\Delta U = \frac{\pi^2 N}{4\varepsilon_F}k_B^2 T^2 \tag{3.89}$$

and the electronic specific heat can be obtained as $c_V = \frac{\partial U}{\partial T} = \frac{\pi^2 N}{2\varepsilon_F}k_B^2 T$, which indeed matches the low-temperature experimental data which states that $c_V \propto T$.

The main reason that the quantum explanation is correct is because of the Pauli exclusion principle. This means that only electrons near the Fermi level are thermally excited. Therefore, only a small number of electrons are involved in 'storing heat'. The classical equipartition theorem assumes that all the electrons are involved equally; thus, it gives a specific heat capacity that is far too large.

3.2 Electronic band theory

3.2.1 General properties of electrons in a periodic potential

We now improve our approximation over the Sommerfeld model by
  • including frozen lattice ions, which are represented by a periodic potential
  • continuing to assume that all the electrons do not interact with each other

In short, this is effectively the problem of one electron interacting with a periodic potential.

The (3D) time-independent Schrödinger equation is

$$\left(-\frac{\hbar^2}{2m}\vec{\nabla}^2 + V(\vec{r})\right)\psi(\vec{r}) = E\psi(\vec{r}) \tag{3.90}$$

with the periodic potential $V(\vec{r} + \vec{R}) = V(\vec{r})$, where the potential is the same at every (real) lattice site.

The real space lattice translation vector $\vec{R}$ is

$$\vec{R} = n_1\vec{a}_1 + n_2\vec{a}_2 + n_3\vec{a}_3, \tag{3.91}$$

where n_1, n_2, and n_3 are integers and $\vec{a}_1$, $\vec{a}_2$, and $\vec{a}_3$ are real space primitive translation vectors.

We shall now extract two important properties.

3.2.1.1 The first consequence of translational symmetry: Bloch's theorem
We want to show Bloch's theorem, which states that due to translational symmetry, the wave function of the electrons takes a certain form called the 'Bloch's function'.

Proof:

Define a unitary translation operator $\hat{T}_{\vec{R}}$ that has this action on the wave function of the electron:

$$\hat{T}_{\vec{R}}\psi(\vec{r}) = \psi(\vec{r} + \vec{R}) \tag{3.92}$$

Physically, we suspect that the physics should be invariant under translations, so the wave functions at $\vec{r}$ and at $\vec{r} + \vec{R}$ must only differ in terms of phase:

$$\hat{T}_{\vec{R}}\psi(\vec{r}) = \psi(\vec{r} + \vec{R}) = e^{i\alpha}\psi(\vec{r}). \tag{3.93}$$

To get more details about the phase factor α, we recall that $\vec{R} = n_1\vec{a}_1 + n_2\vec{a}_2 + n_3\vec{a}_3$.[10]
Let translation in one integer step in the $\vec{a}_i$ direction be

$$\hat{T}_{\vec{a}_i}\psi(\vec{r}) = \psi(\vec{r} + \vec{a}_i) = e^{i\alpha_i}\psi(\vec{r}) \tag{3.94}$$

so, $\qquad \hat{T}_{\vec{R}}\psi(\vec{r}) = \underbrace{\hat{T}_{\vec{a}_3}\cdots\hat{T}_{\vec{a}_3}}_{n_3\text{ steps}}\underbrace{\hat{T}_{\vec{a}_2}\cdots\hat{T}_{\vec{a}_2}}_{n_2\text{ steps}}\underbrace{\hat{T}_{\vec{a}_1}\cdots\hat{T}_{\vec{a}_1}}_{n_1\text{ steps}}\psi(\vec{r}) \tag{3.95}$

$$\psi(\vec{r} + \vec{R}) = \hat{T}_{\vec{a}_3}\cdots\hat{T}_{\vec{a}_3}\hat{T}_{\vec{a}_2}\cdots\hat{T}_{\vec{a}_2}e^{in_1\alpha_1}\psi(\vec{r}) \tag{3.96}$$

$$\psi(\vec{r} + \vec{R}) = e^{i(n_1\alpha_1+n_2\alpha_2+n_3\alpha_3)}\psi(\vec{r}), \tag{3.97}$$
| which we can thus write as a dot product

---

[10] We are not proving Bloch's theorem rigorously. The rigorous proof makes use of group theory.

$$\psi\left(\vec{r} + \vec{R}\right) = e^{i\vec{k}\cdot\vec{R}}\psi(\vec{r}). \tag{3.98}$$

Here, $\vec{k}$ is defined to be $\vec{k} = \frac{\alpha_1}{2\pi}\vec{b}_1 + \frac{\alpha_2}{2\pi}\vec{b}_2 + \frac{\alpha_3}{2\pi}\vec{b}_3$, which is a reciprocal space vector because it is a vector of inverse length units. The reciprocal space primitive translation vectors are the usual (recall equation (1.1)):

$$\vec{b}_1 = 2\pi\frac{\vec{a}_2 \times \vec{a}_3}{|\vec{a}_1 \cdot \vec{a}_2 \times \vec{a}_3|}, \qquad \vec{b}_2 = 2\pi\frac{\vec{a}_3 \times \vec{a}_1}{|\vec{a}_1 \cdot \vec{a}_2 \times \vec{a}_3|}, \qquad \vec{b}_3 = 2\pi\frac{\vec{a}_1 \times \vec{a}_2}{|\vec{a}_1 \cdot \vec{a}_2 \times \vec{a}_3|} \tag{3.99}$$

where $\vec{a}_i \cdot \vec{b}_j = 2\pi\delta_{ij}$.

If we write the wave function function in the form $\psi(\vec{r}) = e^{i\vec{k}\cdot\vec{r}}u_{\vec{k}}(\vec{r})$, then

$$\psi(\vec{r} + \vec{R}) = e^{i\vec{k}\cdot\vec{R}}\psi(\vec{r}) \tag{3.100}$$

$$\implies e^{i\vec{k}\cdot(\vec{r}+\vec{R})}u_{\vec{k}}(\vec{r} + \vec{R}) = e^{i\vec{k}\cdot\vec{R}}e^{i\vec{k}\cdot\vec{r}}u_{\vec{k}}(\vec{r}) \tag{3.101}$$

$$\implies u_{\vec{k}}(\vec{r} + \vec{R}) = u_{\vec{k}}(\vec{r}); \tag{3.102}$$

thus the so-called cell function $u_{\vec{k}}(\vec{r})$ is invariant under translations.

To summarise:[11]

Bloch's theorem: $\psi(\vec{r} + \vec{R}) = e^{i\vec{k}\cdot\vec{R}}\psi(\vec{r})$ | The Bloch function: $\psi(\vec{r}) = e^{i\vec{k}\cdot\vec{r}}u_{\vec{k}}(\vec{r})$

A simple interpretation of $\psi(\vec{r})$ is that $e^{i\vec{k}\cdot\vec{r}}$ is a 3D plane wave and $u_{\vec{k}}(\vec{r})$ is a periodic function that modulates the plane wave.

3.2.1.2 The second consequence of translational symmetry: reduction to the first Brillouin zone

We want to show that the wave vector $\vec{k}$ labelling the Bloch function $\psi(\vec{r})$ is not unique. All Bloch functions with $\vec{k}'$ that are related to $\vec{k}$ by $\vec{k}' = \vec{k} \pm \vec{G}$ are equivalent. Recall that $\vec{G}$ is the reciprocal space translation vector, $\vec{G} = h\vec{b}_1 + k\vec{b}_2 + l\vec{b}_3$, where h, k, and l are integers.

Recall that $\quad \psi(\vec{r}) = e^{i\vec{k}\cdot\vec{r}}u_{\vec{k}}(\vec{r}) \quad$ and $\quad \psi\left(\vec{r} + \vec{R}\right) = e^{i\vec{k}\cdot\vec{R}}\psi(\vec{r}).$ (3.103)

---

[11] Since $\psi(\vec{r}) = e^{i\vec{k}\cdot\vec{r}}u_{\vec{k}}(\vec{r})$, this means $\psi(\vec{r})$ can be secretly labelled by $\vec{k}$. In many books, the notation is then changed from $\psi(\vec{r})$ to $\psi_{\vec{k}}(\vec{r})$ to show this label. In these notes, I am not switching notation.

Now consider a Bloch function labelled by $\vec{k}'$:

$$\psi'(\vec{r}) = e^{i\vec{k}'\cdot\vec{r}}u_{\vec{k}'}(\vec{r})$$

$$\text{| assume that } \vec{k}' = \vec{k} \pm \vec{G}$$

$$(3.104)$$

$$\psi'(\vec{r}) = e^{i(\vec{k}\pm\vec{G})\cdot\vec{r}}u_{\vec{k}\pm\vec{G}}(\vec{r}) \tag{3.105}$$

under a translation, $\quad \psi'(\vec{r} + \vec{R}) = e^{i(\vec{k}\pm\vec{G})\cdot(\vec{r}+\vec{R})}u_{\vec{k}\pm\vec{G}}(\vec{r} + \vec{R}) \tag{3.106}$

$$\psi'(\vec{r} + \vec{R}) = e^{i\vec{k}\cdot\vec{R}}e^{\pm i\vec{G}\cdot\vec{R}}\psi'(\vec{r})$$

$$\text{| recall that } e^{\pm i\vec{G}\cdot\vec{R}} = e^{\pm i2\pi \times \text{integer}} = 1$$

$$(3.107)$$

$$\psi'(\vec{r} + \vec{R}) = e^{i\vec{k}\cdot\vec{R}}\psi'(\vec{r}) \tag{3.108}$$

The appearance of the factor $e^{i\vec{k}\cdot\vec{R}}$ upon translation means that, under translation, ψ' (which is labelled by $\vec{k}'$) can really be labelled by $\vec{k}$. So ψ' and ψ (which is labelled by $\vec{k}$) are equivalent provided that they are related by a reciprocal translation vector $\vec{G}$.

Conclusion:

The complete set of Bloch functions can thus be described by all the $\vec{k}$s in a primitive cell of the reciprocal lattice. Recall that it is usual to use the Wigner–Seitz method to construct the primitive cell in reciprocal space. This cell is called the Brillouin zone. Therefore, we usually just need all the $\vec{k}$s in the first Brillouin zone (1BZ).

Note that we already have hints that 1BZ contains all the physics from chapter 2 but the proof here is strong.

3.2.1.3 Remarks about $\vec{k}$

For a free electron, $\vec{p} = \hbar\vec{k}$ is the momentum of the electron. For an electron in a periodic potential, there are two reasons which show that $\hbar\vec{k}$ is not simply the momentum.

1. $\vec{k}$ and $\vec{k} \pm \vec{G}$ describe the same physics, as seen in the previous section. If $\vec{k}$ were the momentum, then $\vec{k} \pm \vec{G}$ would be a different momentum with a different physics for the electron.

2. The Bloch functions, which are labelled by $\vec{k}$, are not eigenstates of the momentum operator. We shall show this by testing this expression:

$$\widehat{P}\,|\psi\rangle \overset{?}{=} \vec{p}\,|\psi\rangle \tag{3.109}$$

In the position representation $\widehat{P} \rightarrow \frac{\hbar}{i}\vec{\nabla}$,

$$\frac{\hbar}{i}\vec{\nabla}\psi(\vec{r}) = \frac{\hbar}{i}\vec{\nabla}\left(e^{i\vec{k}\cdot\vec{r}}u_{\vec{k}}(\vec{r})\right) \tag{3.110}$$

$$\frac{\hbar}{i}\vec{\nabla}\psi(\vec{r}) = \hbar\vec{k}\psi(\vec{r}) + e^{i\vec{k}\cdot\vec{r}}\left(\frac{\hbar}{i}\vec{\nabla}u_{\vec{k}}(\vec{r})\right) \tag{3.111}$$

$$\frac{\hbar}{i}\vec{\nabla}\psi(\vec{r}) \neq \hbar\vec{k}\psi(\vec{r}) \tag{3.112}$$

Thus we shall call $\hbar\vec{k}$ the 'crystal momentum' to distinguish it from real momentum. The other common name is 'quasimomentum'.

3.2.1.4 Born–von Karman periodic boundary conditions

Actually, for translational symmetry to hold strictly, we would require the solid to be infinite so that there is no surface. Otherwise the translational symmetry will 'end' due to the surface. The properties on the surface are also different from those in the bulk.

In reality, a solid is finite, which we can describe as follows: there are N_1 atoms in direction $\vec{a}_1$, N_2 atoms in direction $\vec{a}_2$, and N_3 atoms in direction $\vec{a}_3$. We want translational symmetry to hold; how can this be achieved?

It is done by imposing Born–von Karman periodic boundary conditions:[12]

$$\psi(\vec{r} + N_i\vec{a}_i) = \psi(\vec{r}) \quad \text{with } i = 1, 2, 3 \tag{3.113}$$

Applying these to the Bloch function form, we obtain:

$$e^{i\vec{k}\cdot(\vec{r}+N_i\vec{a}_i)}u_{\vec{k}}(\vec{r} + N_i\vec{a}_i) = e^{i\vec{k}\cdot\vec{r}}u_{\vec{k}}(\vec{r})$$
$$| \text{ recall that } u_{\vec{k}}(\vec{r} + N_i\vec{a}_i) = u_{\vec{k}}(\vec{r}) \tag{3.114}$$

$$\implies e^{i\vec{k}\cdot(N_i\vec{a}_i)} = 1 \tag{3.115}$$

$$\implies N_i\vec{k} \cdot \vec{a}_i = 2\pi n \quad \text{where } n \text{ is an integer}$$
$$| \text{ recall that } \vec{k} = \frac{\alpha_1}{2\pi}\vec{b}_1 + \frac{\alpha_2}{2\pi}\vec{b}_2 + \frac{\alpha_3}{2\pi}\vec{b}_3 \text{ and } \vec{a}_i \cdot \vec{b}_j = 2\pi\delta_{ij} \tag{3.116}$$

$$N_i\frac{\alpha_i}{2\pi}2\pi = 2\pi n \tag{3.117}$$

$$\frac{\alpha_i}{2\pi} = \frac{n}{N_i} \tag{3.118}$$

---

[12] This is equivalent to joining the last atom to the origin. It is, of course, unrealistic and is meant to be a theoretical simplification.

This means that the allowed wave vectors are

$$\vec{k} = \frac{n}{N_1}\vec{b}_1 + \frac{m}{N_2}\vec{b}_2 + \frac{p}{N_3}\vec{b}_3 \qquad \text{where } n, m, \text{ and } p \text{ are integers.} \qquad (3.119)$$

- Thus, there are $N_1 \times N_2 \times N_3$ allowed $\vec{k}$ points in the 1BZ. To see this, take the case of 1D; the 1BZ boundaries are at $-\frac{1}{2}\vec{b}_1$ and $\frac{1}{2}\vec{b}_1$. This means that at the $\frac{1}{2}\vec{b}_1$ zone boundary, $n = \frac{N_1}{2}$. Thus, across the 1BZ from $-\frac{1}{2}\vec{b}_1$ to $\frac{1}{2}\vec{b}_1$, we have N_1 allowed $\vec{k}$ points.
- This leads to the fact that the 'number of allowed $\vec{k}$ points in the 1BZ = the number of real space primitive cells'.
- Also,

$$\text{the volume of one } \vec{k}\text{-point} = \frac{1}{N_1}\vec{b}_1 \cdot \frac{1}{N_2}\vec{b}_2 \times \frac{1}{N_3}\vec{b}_3; \qquad (3.120)$$

$$\text{the volume of one } \vec{k} - \text{point} = \frac{\vec{b}_1 \cdot \vec{b}_2 \times \vec{b}_3}{N_1 N_2 N_3}; \qquad (3.121)$$

$$\text{the volume of one } \vec{k} - \text{point} = \frac{\text{the volume of the reciprocal space primitive cell}}{N_1 N_2 N_3}. \qquad (3.122)$$

3.2.1.5 The central equation

We will now derive an equation called the 'central equation'. It is simply the Schrödinger equation for a particle in a periodic potential.[13]

Recall that the potential is periodic in $\vec{R}: V(\vec{r}) = V(\vec{r} + \vec{R})$; thus, it can be written into the following Fourier series:

$$V(\vec{r}) = \sum_{\vec{G}} V_{\vec{G}} e^{i\vec{G}\cdot\vec{r}}. \qquad (3.123)$$

The wave function has no special periodicity, so its Fourier series is just a generic sum of plane waves:

$$\psi(\vec{r}) = \sum_{\vec{k}} C_{\vec{k}} e^{i\vec{k}\cdot\vec{r}}. \qquad (3.124)$$

If we have a finite number of ions and we want to avoid surface effects, then $\vec{k}$ must be subjected to the Born–von Karman periodic boundary conditions. The

---

[13] When the solid is infinite, the allowed $\vec{k}$ is dense in the 1BZ. This conclusion has already been encountered in chapter 2.

central equation is the equation that solves for the coefficients $C_{\vec{k}}$, and hence the Bloch function $\psi(\vec{r})$ is determined.

Substitute $V(\vec{r})$ and $\psi(\vec{r})$ into the one-particle Schrödinger equation with a periodic potential:

$$\left(-\frac{\hbar^2}{2m}\vec{\nabla}^2 + V(\vec{r})\right)\psi(\vec{r}) = E\psi(\vec{r}) \tag{3.125}$$

$$-\frac{\hbar^2}{2m}\vec{\nabla}^2\sum_{\vec{k}}C_{\vec{k}}e^{i\vec{k}\cdot\vec{r}} + \sum_{\vec{G}}V_{\vec{G}}e^{i\vec{G}\cdot\vec{r}}\sum_{\vec{k}}C_{\vec{k}}e^{i\vec{k}\cdot\vec{r}} = E\sum_{\vec{k}}C_{\vec{k}}e^{i\vec{k}\cdot\vec{r}} \tag{3.126}$$

note that $\vec{\nabla}^2 e^{i\vec{k}\cdot\vec{r}} = -k^2 e^{i\vec{k}\cdot\vec{r}}$ |

$$\sum_{\vec{k}}\left(\frac{\hbar^2 k^2}{2m} - E\right)C_{\vec{k}}e^{i\vec{k}\cdot\vec{r}} + \sum_{\vec{G}}\sum_{\vec{k}}V_{\vec{G}}C_{\vec{k}}e^{i(\vec{G}+\vec{k})\cdot\vec{r}} = 0$$

rename $\vec{G} + \vec{k} = \vec{k}'$, so $\sum_{\vec{G}}\sum_{\vec{k}}V_{\vec{G}}C_{\vec{k}}e^{i(\vec{G}+\vec{k})\cdot\vec{r}} = \sum_{\vec{G}}\sum_{\vec{k}}V_{\vec{G}}C_{\vec{k}'-\vec{G}}e^{i\vec{k}'\cdot\vec{r}}$ |

then recall $\vec{k} = \frac{n}{N_1}\vec{b}_1 + \frac{m}{N_2}\vec{b}_2 + \frac{p}{N_3}\vec{b}_3$ and $\vec{G} = h\vec{b}_1 + k\vec{b}_2 + l\vec{b}_3$ |

means $\vec{k}' = \left(\frac{n}{N_1} + h\right)\vec{b}_1 + \left(\frac{m}{N_2} + k\right)\vec{b}_2 + \left(\frac{p}{N_3} + l\right)\vec{b}_3$ | $\qquad$ (3.127)

so summing over all $\vec{k}$ and $\vec{G}$ is the same as over all $\vec{k}'$ and $\vec{G}$ |

thus, $\sum_{\vec{G}}\sum_{\vec{k}}V_{\vec{G}}C_{\vec{k}'-\vec{G}}e^{i\vec{k}'\cdot\vec{r}} = \sum_{\vec{G}}\sum_{\vec{k}'}V_{\vec{G}}C_{\vec{k}'-\vec{G}}e^{i\vec{k}'\cdot\vec{r}}$ |

then relabel $\vec{k}'$ as $\vec{k}$, so $\sum_{\vec{G}}\sum_{\vec{k}'}V_{\vec{G}}C_{\vec{k}'-\vec{G}}e^{i\vec{k}'\cdot\vec{r}} \; \sum_{\vec{G}}\sum_{\vec{k}}V_{\vec{G}}C_{\vec{k}-\vec{G}}e^{i\vec{k}\cdot\vec{r}}$ |

$$\sum_{\vec{k}}\left\{\left(\frac{\hbar^2 k^2}{2m} - E\right)C_{\vec{k}} + \sum_{\vec{G}}V_{\vec{G}}C_{\vec{k}-\vec{G}}\right\}e^{i\vec{k}\cdot\vec{r}} = 0 \tag{3.128}$$

since the plane waves $e^{i\vec{k}\cdot\vec{r}}$ are linearly independent, thus |

$$\left(\frac{\hbar^2 k^2}{2m} - E\right)C_{\vec{k}} + \sum_{\vec{G}}V_{\vec{G}}C_{\vec{k}-\vec{G}} = 0. \tag{3.129}$$

This is the central equation, which is simply the Schrödinger equation for a particle in a periodic potential in terms of Fourier coefficients.

As we have learnt that the crystal momentum in the 1BZ contains all the essential physics, it is therefore convenient to write the central equation in terms of wave vectors belonging to the 1BZ. We shall label them using $\vec{q}$. So $\vec{k} = \vec{q} - \vec{G}'$, where $\vec{G}'$ is a suitable reciprocal translation vector to map $\vec{k}$ to $\vec{q}$ in the 1BZ. Thus,

$$\left(\frac{\hbar^2(\vec{q} - \vec{G}')^2}{2m} - E\right)C_{\vec{q}-\vec{G}'} + \sum_{\vec{G}}V_{\vec{G}}C_{\vec{q}-\vec{G}'-\vec{G}} = 0 \qquad (3.130)$$

relabel $\vec{G}' + \vec{G} = \vec{G}''$ and note that summing all $\vec{G}$ and summing all $\vec{G}''$ is the same. |

$$\boxed{\left(\frac{\hbar^2(\vec{q} - \vec{G}')^2}{2m} - E\right)C_{\vec{q}-\vec{G}'} + \sum_{\vec{G}''}V_{\vec{G}''-\vec{G}'}C_{\vec{q}-\vec{G}''} = 0} \qquad (3.131)$$

This is another form of the central equation expressed in terms of wave vectors $\vec{q}$ which are within the 1BZ.[14]

We now need to see whether there is any contradiction between the Bloch function $\psi(\vec{r}) = e^{i\vec{q}\cdot\vec{r}}u_{\vec{q}}(\vec{r})$, whose 1BZ crystal momentum label is $\vec{q}$, and the Fourier series form introduced earlier: $\psi(\vec{r}) = \sum_{\vec{k}}C_{\vec{k}}e^{i\vec{k}\cdot\vec{r}}$.

Start with $\quad \psi(\vec{r}) = \sum_{\vec{k}}C_{\vec{k}}e^{i\vec{k}\cdot\vec{r}};$

$\quad$ | recall that $\vec{k} = \vec{q} - \vec{G}'$ with $\vec{q}$ in the 1BZ. $\qquad (3.132)$

$\quad$ | Since each Bloch function is labelled by a fixed $\vec{k}$, so $\vec{q}$ is fixed.

$\quad$ | The sum now runs over $\vec{G}'$ to reach all 'equivalent' $\vec{k}$s.

$$\psi(\vec{r}) = \sum_{\vec{G}'}C_{\vec{q}-\vec{G}'}e^{i(\vec{q}-\vec{G}')\cdot\vec{r}} \qquad (3.133)$$

$$\psi(\vec{r}) = e^{i\vec{q}\cdot\vec{r}}\left(\sum_{\vec{G}'}C_{\vec{q}-\vec{G}'}e^{-i\vec{G}'\cdot\vec{r}}\right) \qquad (3.134)$$

This looks like the Bloch form, so can we call $\sum_{\vec{G}'}C_{\vec{q}-\vec{G}'}e^{-i\vec{G}'\cdot\vec{r}} = u_{\vec{q}}(\vec{r})$, the cell function? In other words, does this expression for $u_{\vec{q}}(\vec{r})$ have the right periodic property: $u_{\vec{q}}(\vec{r} + \vec{R}) = u_{\vec{q}}(\vec{r})$? Let us try

$$u_{\vec{q}}(\vec{r} + \vec{R}) = \sum_{\vec{G}'}C_{\vec{q}-\vec{G}'}e^{-i\vec{G}'\cdot(\vec{r}+\vec{R})} \qquad (3.135)$$

$\quad$ | recall that $e^{-i\vec{G}'\cdot\vec{R}} = 1$

$$u_{\vec{q}}(\vec{r} + \vec{R}) = \sum_{\vec{G}'}C_{\vec{q}-\vec{G}'}e^{-i\vec{G}'\cdot\vec{r}} \qquad (3.136)$$

indeed, $u_{\vec{q}}(\vec{r} + \vec{R}) = u_{\vec{q}}(\vec{r})$. $\qquad (3.137)$

Thus, there is no contradiction.

---

[14] Don't forget the meanings: $\vec{G}'$ is a suitable reciprocal translation vector to keep the wave vector within the 1BZ and $\vec{G}''$ is a dummy reciprocal translation vector.

3.2.2 Models for band structure calculations

3.2.2.1 The nearly free electron model

[1D Model] We want to make use of the central equation to extract the physics of how the electron behaves in a periodic potential. For simplicity, we start with a 1D model.

The real space consists of ions that have the spacing a, and thus the reciprocal space has points separated by $\frac{2\pi}{a}$ (figure 3.6). From the central equation, the most drastic approximation we can make is to assume that there is no periodic potential $V(\vec{r}) = 0$ (and so all its Fourier coefficients are zero), so

$$\left(\frac{\hbar^2(\vec{q} - \vec{G}')^2}{2m} - E \right) C_{\vec{q}-\vec{G}'} + \sum_{\vec{G}''} \underbrace{V_{\vec{G}''-\vec{G}'}}_{=0} C_{\vec{q}-\vec{G}''} = 0 \tag{3.138}$$

the wave function coefficients $C_{\vec{q}-\vec{G}'}$ should not be zero |

$$E = \frac{\hbar^2(\vec{q} - \vec{G}')^2}{2m}. \tag{3.139}$$

This is actually called the empty lattice approximation. The dispersion relation[15] $E = \frac{\hbar^2(\vec{q} - \vec{G}')^2}{2m}$ represents that of a free electron, which is parabolic. However, there is k-space periodicity in this dispersion relation even though there is no ion lattice. See the repeated parabolas in figure 3.7.

Next, we proceed with the nearly free electron model, which means that the periodic potential $V(\vec{r})$ is very weak and small. We shall work out the effects of this weak potential in the region of two degenerate levels. See figure 3.7 for a simple reason why this region is relevant.

We start by writing the central equation for each level:[16]

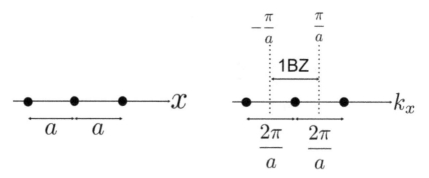

Figure 3.6. LEFT: a 1D real space lattice that has the spacing a. RIGHT: the corresponding reciprocal space with the 1BZ marked.

---

[15] Let me remind you that the dispersion relation is the relation of E versus $\vec{k}$ or ω versus $\vec{k}$, where $\omega = \frac{E}{\hbar}$.
[16] Note that these two levels are one reciprocal translation vector or more away from each other.

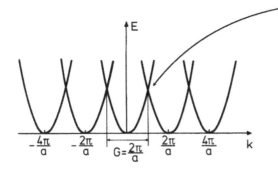

Region of 2 degenerate free electron levels where the effect of the (weak) potential will be most significant.

Simple justification: the difference in energy here is the same order as V.

Figure 3.7. The effect of the weak periodic potential is linear in $V_{\vec{G}}$ near these degenerate points, while at non-degenerate energies, the effect is quadratic in $V_{\vec{G}}$. This means that the effect of the weak periodic potential (where $V_{\vec{G}}$ is small) is most pronounced around the degenerate levels. For justification, see [6] page 155.

$$\text{level} \frac{\hbar^2\left(\vec{q}-\vec{G}_a'\right)^2}{2m}:$$

$$\left(\frac{\hbar^2\left(\vec{q}-\vec{G}_a'\right)^2}{2m} - E\right) C_{\vec{q}-\vec{G}_a'} + V_{\vec{G}_a'-\vec{G}_a'} C_{\vec{q}-\vec{G}_a'} + V_{\vec{G}_b'-\vec{G}_a'} C_{\vec{q}-\vec{G}_b'} + \sum_{\vec{G}''\,(\neq \vec{G}_a',\vec{G}_b')} V_{\vec{G}''-\vec{G}_a'} C_{\vec{q}-\vec{G}''} = 0$$

$$\text{level} \frac{\hbar^2\left(\vec{q}-\vec{G}_b'\right)^2}{2m}:$$

$$\left(\frac{\hbar^2\left(\vec{q}-\vec{G}_b'\right)^2}{2m} - E\right) C_{\vec{q}-\vec{G}_b'} + V_{\vec{G}_b'-\vec{G}_b'} C_{\vec{q}-\vec{G}_b'} + V_{\vec{G}_a'-\vec{G}_b'} C_{\vec{q}-\vec{G}_a'} + \sum_{\vec{G}''\,(\neq \vec{G}_b',\vec{G}_a')} V_{\vec{G}''-\vec{G}_b'} C_{\vec{q}-\vec{G}''} = 0$$

- Note that $V_{\vec{G}_a'-\vec{G}_a'} = V_{\vec{G}_b'-\vec{G}_b'} = V_0$, which is the first term in the Fourier series $V(\vec{r}) = \sum_{\vec{G}} V_{\vec{G}} e^{i\vec{G}\cdot\vec{r}}$. This first term is a constant and energy can be shifted by a constant, so we set $V_0 = 0$.
- Note that each level corresponds to a plane wave; say, level $\frac{\hbar^2(\vec{q}-\vec{G}_b')^2}{2m}$ corresponds to $\psi \propto e^{i(\vec{q}-\vec{G}_b')\cdot\vec{r}}$, so the coefficient $C_{\vec{q}-\vec{G}_b'}$ dominates. The other coefficients are small, so we drop the last term of each central equation.
- We label $\frac{\hbar^2(\vec{q}-\vec{G}_a')^2}{2m} = E^{\text{free}}_{\vec{q}-\vec{G}_a'}$ and $\frac{\hbar^2(\vec{q}-\vec{G}_b')^2}{2m} = E^{\text{free}}_{\vec{q}-\vec{G}_b'}$ for ease of writing.

$$\text{level} \frac{\hbar^2(\vec{q}-\vec{G}_a')^2}{2m}: \qquad \left(E^{\text{free}}_{\vec{q}-\vec{G}_a'} - E\right) C_{\vec{q}-\vec{G}_a'} + V_{\vec{G}_b'-\vec{G}_a'} C_{\vec{q}-\vec{G}_b'} = 0 \qquad (3.140)$$

$$\text{level } \frac{\hbar^2(\vec{q} - \vec{G}_b')^2}{2m}: \quad \left(E^{\text{free}}_{\vec{q}-\vec{G}'_b} - E \right) C_{\vec{q}-\vec{G}'_b} + V_{\vec{G}'_a-\vec{G}'_b} C_{\vec{q}-\vec{G}'_a} = 0 \quad (3.141)$$

- By now, it is clear that any new physics appearing at the end is due to the superposition of these two plane waves.
- Note that as the potential $V(\vec{r})$ is real,

$$V^*(\vec{r}) = V(\vec{r}) \implies \sum_{\vec{G}} V^*_{\vec{G}} e^{-i\vec{G}\cdot\vec{r}} = \sum_{\vec{G}} V_{\vec{G}} e^{i\vec{G}\cdot\vec{r}} \quad (3.142)$$

|then rename $\vec{G} \longrightarrow -\vec{G}$ on the right−hand side

$$\sum_{\vec{G}} V^*_{\vec{G}} e^{-i\vec{G}\cdot\vec{r}} = \sum_{\vec{G}} V_{-\vec{G}} e^{-i\vec{G}\cdot\vec{r}} \quad (3.143)$$

$$\implies V^*_{\vec{G}} = V_{-\vec{G}} \quad (3.144)$$

$$\implies V^*_{\vec{G}'_a-\vec{G}'_b} = V_{\vec{G}'_b-\vec{G}'_a}. \quad (3.145)$$

From the coupled equations (3.140) and (3.141), the solution for E is found by solving det (coefficient matrix) = 0,

$$E = \frac{1}{2}\left(E^{\text{free}}_{\vec{q}-\vec{G}'_a} + E^{\text{free}}_{\vec{q}-\vec{G}'_b} \right) \pm \sqrt{\left(\frac{E^{\text{free}}_{\vec{q}-\vec{G}'_a} - E^{\text{free}}_{\vec{q}-\vec{G}'_b}}{2} \right)^2 + \left| V_{\vec{G}'_a-\vec{G}'_b} \right|^2} \quad (3.146)$$

| at the BZ boundary, the two E^{free}s are equal, so the first term in the square root vanishes

$$E = \frac{1}{2}\left(E^{\text{free}}_{\vec{q}-\vec{G}'_a} + E^{\text{free}}_{\vec{q}-\vec{G}'_b} \right) \pm \left| V_{\vec{G}'_a-\vec{G}'_b} \right| \quad (3.147)$$

So at the BZ boundary, a symmetrical split of levels occurs. See figures 3.8 and 3.9. There is thus an energy gap, which is called the bandgap. This physical effect is real and it will be shown to emerge from other methods of calculation, namely the tight-binding model and the exact Kronig–Penney model.

We shall now look at three schemes in which the dispersion relation can be represented in figure 3.8.

We shall push this discussion further and rediscover a nice relationship. Recall that at the BZ boundary, $E^{\text{free}}_{\vec{q}-\vec{G}'_a} = E^{\text{free}}_{\vec{q}-\vec{G}'_b}$, so

$$|\vec{q} - \vec{G}'_a| = |\vec{q} - \vec{G}'_b| \quad (3.148)$$

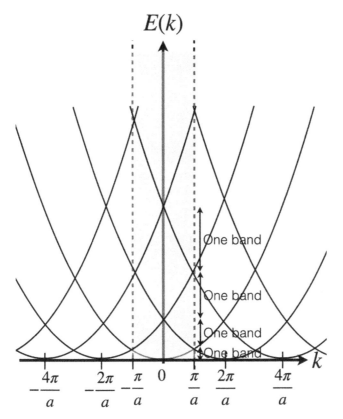

Figure 3.8. The three schemes are: the extended zone scheme, the reduced zone scheme, and the repeated zone scheme. This diagram illustrates the extended zone scheme in the empty lattice approximation. The reduced zone scheme is simply the dispersion within the 1BZ shaded in yellow. The repeated zone scheme is just the reduced zone scheme repeated in every BZ, so it is highly redundant.

$$\begin{aligned} &| \text{write } \vec{G}_- = \vec{G}'_b - \vec{G}'_a \text{ and } \vec{k} = \vec{q} - \vec{G}'_a \\ &|\vec{k}| = |\vec{k} - \vec{G}_-| \end{aligned} \tag{3.149}$$

Graphically, this looks like figure 3.10 because $\vec{k}$ and $\vec{k} - \vec{G}_-$ have the same magnitude.

Thus the vector $\vec{k} - \frac{1}{2}\vec{G}_-$ lies in the bisector plane and $\vec{G}_- \cdot \left(\vec{k} - \frac{1}{2}\vec{G}_-\right) = 0$ $\implies |\vec{G}_-|^2 = 2\vec{k} \cdot \vec{G}_-$. This is the Bragg condition![17]
[2D Model]

We can consider a 2D square lattice with atomic spacing a. The real space translation vector is $\vec{R} = n_1\vec{a}_1 + n_2\vec{a}_2$, where $\vec{a}_1 = a\hat{x}$ and $\vec{a}_2 = a\hat{y}$. The reciprocal

---

[17] We should not think of the electron wave as undergoing Bragg diffraction. The electron wave is affected before the BZ boundary, where x-ray/neutron diffraction only happens when the Bragg condition is exactly met.

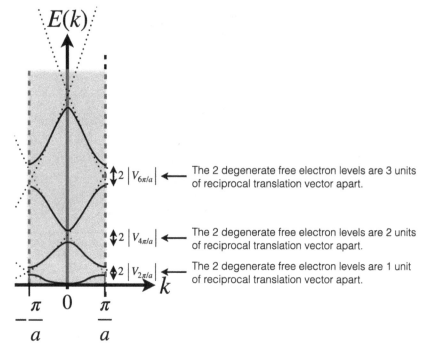

The 2 degenerate free electron levels are 3 units of reciprocal translation vector apart.

The 2 degenerate free electron levels are 2 units of reciprocal translation vector apart.

The 2 degenerate free electron levels are 1 unit of reciprocal translation vector apart.

Figure 3.9. The size of the bandgap is twice the Fourier coefficient of the potential. The relevant Fourier coefficient is determined by the number of reciprocal translations that separate the two degenerate states. Animation available at https://doi.org/10.1088/978-0-7503-5217-8.

Bisector Plane

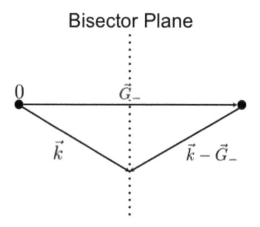

Figure 3.10. The Bragg condition is met by the electron wave at the BZ boundary.

space is square and has the spacing $\frac{2\pi}{a}$. The reciprocal space translation vector is $\vec{G} = h\vec{b}_1 + k\vec{b}_2$, where $\vec{b}_1 = \frac{2\pi}{a}\hat{x}$ and $\vec{b}_2 = \frac{2\pi}{a}\hat{y}$. The high symmetry points in the 1BZ are shown and labelled in figure 3.11.

The dispersion relation in the empty lattice approximation is a paraboloid centred at every k-point (points at the left of figure 3.11, not points at the bottom right of

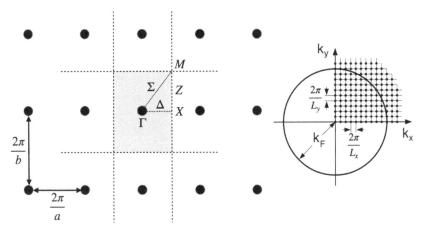

Figure 3.11. LEFT: the 2D reciprocal space of the 2D real space square lattice. The first three BZs are shown in different colours. The high symmetry points and directions in the 1BZ are labelled. RIGHT: A 2D Fermi sphere (whose size depends on electron density) and allowed k-points as required by Born–von Karman periodic boundary conditions. Animation available at https://doi.org/10.1088/978-0-7503-5217-8.

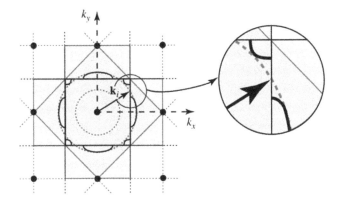

Figure 3.12. The 'rounding out' of the 2D Fermi sphere to meet the 1BZ boundary perpendicularly. Reprinted by permission from Springer Nature [1], Copyright (2010).

figure 3.11). The constant energy 'surface' is a circle. The Fermi 'sphere' is therefore the largest circle based on the electron density.

We are going to assume a weak potential again, and so the effect of the potential is again going to be the most pronounced where two paraboloids touch, which is at the BZ boundary. The effect is similar to that in 1D: the Fermi 'sphere' will 'round out' and meet the BZ boundary perpendicularly (figure 3.12).

[3D Model]

For a 3D solid, a full drawing of the dispersion relation would require a 4D drawing (E, k_x, k_y, k_z), which is impossible, so we can only draw cross sections. See figure 3.23, for example.

We can already guess the effect of a weak potential, which is to 'round out' the Fermi surface to meet the BZ boundaries (figure 3.13 and 3.14). So we shall not repeat the story again; let us simply go to specific cases of real metals:

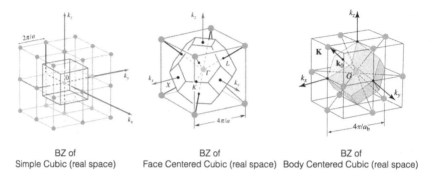

| BZ of | BZ of | BZ of |
| Simple Cubic (real space) | Face Centered Cubic (real space) | Body Centered Cubic (real space) |

Figure 3.13. Here, we put together the 1BZs of three common cubic structures. Reprinted by permission from Springer Nature [1], Copyright (2010).

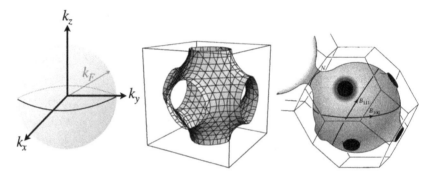

Figure 3.14. LEFT: The Fermi sphere for 3D free electrons. Its size depends on the electron density. If the Fermi sphere meets the BZ boundaries, we get the 'rounding out' effect. MIDDLE: 'Rounding out' by a cubic BZ boundary. Reproduced with permission from [1]. RIGHT: the Fermi surface of copper. Reproduced with permission from [2].

- Alkali metals:
    - These contribute one electron per (real space) primitive cell.
    - They have a BCC structure. This means there are two atoms in a conventional cell, so the electron density is $n = \dfrac{2}{a^3}$.
    - The Fermi wave vector is $k_F = (3\pi^2 n)^{1/3} = \dfrac{\left(6\pi^2\right)^{1/3}}{a} = \dfrac{1.24\pi}{a}$.
    - The shortest distance to the BZ boundary is half the length of the reciprocal space vector: $\dfrac{1}{2}\dfrac{2\pi}{a}(1^2 + 1^2 + 0^2)^{1/2} = \dfrac{1.41\pi}{a}$.
    - Thus, the Fermi sphere only reaches $\dfrac{1.24\pi}{a} \div \dfrac{1.41\pi}{a} \times 100\% = 88\%$ of the way to the nearest zone boundary; therefore, it does not really get distorted by the effects of the potential.
    - As a result, alkali metals have electronic properties similar to those predicted by the Sommerfeld model.

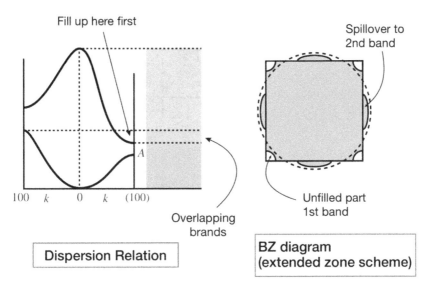

Fill up here first

Spillover to
2nd band

A

100   *k*   0   *k*   (100)

Unfilled part
1st band

Overlapping
brands

Dispersion Relation

**BZ diagram
(extended zone scheme)**

Figure 3.15. Diagrams to show why Group 2 elements may be good conductors of electricity. The band overlap results in partially filled bands. Reprinted with permission from [3], Cambridge University Press.

- Elements with even numbers of valence electrons:
  - We recall that the number of allowed k-points and the number of (real space) primitive cells are the same. Also, each k-point holds two electrons. So for elements with even valence electrons, the electrons can fill bands completely.
  - Thus the topmost electrons are always an energy gap away from the next unoccupied band. This implies that such elements should be insulators (like diamond) or semiconductors (like Ge or Si), depending on the size of the bandgap.
  - Group 2 elements such as Ca have even numbers of valence electrons but they are good conductors because it may be energetically more favourable to fill the next higher band. This results in partially filled bands and thus a conductive solid. See figure 3.15.

3.2.2.2 The tight-binding model
There is another way to describe the formation of bands. This way is somewhat opposite to the nearly free electron model.

The story is that when atoms get closer and closer together, atomic states superpose to form band states. This is the approach of the tight-binding model. In chemistry, this is more commonly known as the 'linear combination of atomic orbitals' (LCAO) method (figure 3.16).

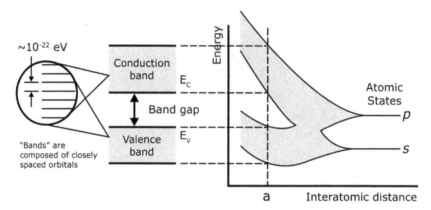

Figure 3.16. The formation of bands in the tight-binding model: atoms are brought closer and overlapping atomic states form bands with bandgaps.

The atomic states are eigenstates of the atomic Hamiltonian[18]

$$H_{atom}\phi_j(\vec{r}) = E_j^{atom}\phi_j(\vec{r}) \quad \text{where } H_{atom} = KE_{atom} + V_{atom}, \tag{3.150}$$

where the index j labels the orbitals of the eigenstate, such as s, p_x, p_y, p_z, d_{xy}.... The bands are formed from the linear combinations of these atomic states at different sites. This is the correct combination (we drop the index j for ease of writing):[19]

$$\psi(\vec{r}) = \sum_{\vec{R}} e^{i\vec{k}\cdot\vec{R}}\phi(\vec{r} - \vec{R}). \tag{3.151}$$

We must check whether it satisfies Bloch's theorem.

$$\psi(\vec{r} + \vec{R}_1) = \sum_{\vec{R}} e^{i\vec{k}\cdot\vec{R}}\phi(\vec{r} + \vec{R}_1 - \vec{R}) \tag{3.152}$$

$$| \text{ let } \vec{R}' = \vec{R} - \vec{R}_1$$

$$\psi(\vec{r} + \vec{R}_1) = \sum_{\vec{R}} e^{i\vec{k}\cdot(\vec{R}'+\vec{R}_1)}\phi(\vec{r} - \vec{R}') \tag{3.153}$$

$$| \text{ summing over } \vec{R} \text{ and } \vec{R}' \text{ is the same}$$

$$\psi(\vec{r} + \vec{R}_1) = e^{i\vec{k}\cdot\vec{R}_1}\sum_{\vec{R}'} e^{i\vec{k}\cdot\vec{R}'}\phi(\vec{r} - \vec{R}') \tag{3.154}$$

$$\psi\left(\vec{r} + \vec{R}_1\right) = e^{i\vec{k}\cdot\vec{R}_1}\psi(\vec{r}), \tag{3.155}$$

which is exactly what Bloch's theorem says.

---

[18] If you recall, according to basic quantum mechanics, H_{atom} is like the Hamiltonian of the hydrogen atom, ϕ are the hydrogen wave functions, and the index j is like the nlm quantum numbers.
[19] This is called the Wannier representation.

We shall now work out the tight-binding model using the simplest example to illustrate the technique. It is the simplest because we are assuming overlapping s-orbital atomic states only; this means that the states are spherically symmetric. We shall take a 3D rectangular real space with primitive translation vectors: $\vec{a}_1 = a\hat{x}$, $\vec{a}_2 = b\hat{y}$, and $\vec{a}_3 = c\hat{z}$.

The (3D, time-independent) Schrödinger equation for the solid is $H\psi(\vec{r}) = \varepsilon_{\vec{k}}\psi(\vec{r})$, where the solid Hamiltonian is

$$H = KE_{\text{atom}} + V(\vec{r}) \tag{3.156}$$

$$\begin{aligned} H &= KE_{\text{atom}} + V_{\text{atom}} + V(\vec{r}) - V_{\text{atom}} \\ &| \text{ call } V(\vec{r}) - V_{\text{atom}} = V' \end{aligned} \tag{3.157}$$

$$H = H_{\text{atom}} + V' \tag{3.158}$$

where $V(\vec{r})$ is the periodic crystal potential. We put everything into the Schrödinger equation.

$$H\psi(\vec{r}) = \varepsilon_{\vec{k}}\psi(\vec{r}) \tag{3.159}$$

$$(H_{\text{atom}} + V')\sum_{\vec{R}}e^{i\vec{k}\cdot\vec{R}}\phi(\vec{r} - \vec{R}) = \varepsilon_{\vec{k}}\sum_{\vec{R}}e^{i\vec{k}\cdot\vec{R}}\phi(\vec{r} - \vec{R}) \tag{3.160}$$

$$| \text{ multiply } \phi^*(\vec{r}) \text{ and integrate } \int d^3\vec{r}$$

$$\int d^3\vec{r}\phi^*(\vec{r})(H_{\text{atom}} + V')\sum_{\vec{R}}e^{i\vec{k}\cdot\vec{R}}\phi(\vec{r} - \vec{R}) = \varepsilon_{\vec{k}}\int d^3\vec{r}\phi^*(\vec{r})\sum_{\vec{R}}e^{i\vec{k}\cdot\vec{R}}\phi(\vec{r} - \vec{R}) \tag{3.161}$$

- LHS first term: $\phi^*H_{\text{atom}} = \phi^*E^{\text{atom}}$, then $\int d^3\vec{r}\phi^*(\vec{r})\phi(\vec{r} - \vec{R}) = 1$ when $\vec{R} \to 0$. This means we have normalisation when the orbitals are at the same site, and we assume that orbitals have vanishing overlap when they are at different sites.
- LHS second term: this is the important term that we will deal with later.
- RHS: gives $\varepsilon_{\vec{k}}$ since $\int d^3\vec{r}\phi^*(\vec{r})\phi(\vec{r} - \vec{R}) = 1$ when $\vec{R} \to 0$. This follows the assumption made earlier.

We can continue with the calculation:

$$\varepsilon_{\vec{k}} = E^{\text{atom}} + \sum_{\vec{R}}e^{i\vec{k}\cdot\vec{R}}\int \phi^*(\vec{r})V'\phi\left(\vec{r} - \vec{R}\right)d^3\vec{r}$$

$$| \text{ assume only nearest neighbours (NNs) contribute,} \tag{3.162}$$
$$| \text{ so the sum runs over } \vec{R} = 0, \vec{R} = \pm a\hat{x}, \vec{R} = \pm b\hat{y}, \vec{R} = \pm c\hat{z}$$
$$| \text{ then label } - \int \phi^*(\vec{r})V'\phi(\vec{r})d^3\vec{r} = E^{\text{onsite}}$$

$$\varepsilon_{\vec{k}} = E^{\text{atom}} - E^{\text{onsite}} + e^{i\vec{k}\cdot a\hat{x}} \int \phi^*(\vec{r})V'\phi(\vec{r} - a\hat{x})d^3\vec{r} + e^{-i\vec{k}\cdot a\hat{x}} \int \phi^*(\vec{r})V'\phi(\vec{r} + a\hat{x})d^3\vec{r}$$

$$+ e^{i\vec{k}\cdot b\hat{y}} \int \phi^*(\vec{r})V'\phi(\vec{r} - b\hat{y})d^3\vec{r} + e^{-i\vec{k}\cdot b\hat{y}} \int \phi^*(\vec{r})V'\phi(\vec{r} + b\hat{y})d^3\vec{r} \qquad (3.163)$$

$$+ e^{i\vec{k}\cdot c\hat{z}} \int \phi^*(\vec{r})V'\phi(\vec{r} - c\hat{z})d^3\vec{r} + e^{-i\vec{k}\cdot c\hat{z}} \int \phi^*(\vec{r})V'\phi(\vec{r} + c\hat{z})d^3\vec{r}$$

$$\varepsilon_{\vec{k}} = E^{\text{atom}} - E^{\text{onsite}} + (e^{ik_x a} + e^{-ik_x a}) \int \phi^*(\vec{r})V'\phi(\vec{r} + a\hat{x})d^3\vec{r}$$

$$(e^{ik_y b} + e^{-ik_y b}) \int \phi^*(\vec{r})V'\phi(\vec{r} + b\hat{y})d^3\vec{r}$$

$$+ (e^{ik_z c} + e^{-ik_z c}) \int \phi^*(\vec{r})V'\phi(\vec{r} + c\hat{z})d^3\vec{r} \qquad (3.164)$$

| so $e^{ik_x a} + e^{-ik_x a} = 2\cos(k_x a)$ and so on

| call $- \int \phi^*(\vec{r})V'\phi(\vec{r} + a\hat{x})d^3\vec{r}$

the x − transfer (or hopping) integral t_x and so on

$$\varepsilon_{\vec{k}} = E^{\text{atom}} - E^{\text{onsite}} - 2t_x \cos(k_x a) - 2t_y \cos(k_y b) - 2t_z \cos(k_z c) \qquad (3.165)$$

This is the required dispersion relation in the tight-binding model. As an aside, note that in the third term of equation (3.163), $\int \phi^*(\vec{r})V'\phi(\vec{r} - a\hat{x})d^3\vec{r}$, relabel $\vec{r} = -\vec{r}'$. Then for atomic wavefunctions that depend on $|\vec{r}|$ only (such as the s-states we assumed here),

$$\int \phi^*(|\vec{r}|)V'\phi(|\vec{r} - a\hat{x}|)d^3\vec{r} = \int \phi^*(|-\vec{r}'|)V'\phi(|-\vec{r}' - a\hat{x}|)d^3\vec{r}'$$

| where $\int d^3\vec{r} = \int d^3\vec{r}'$ because

| if we think Cartesian $\int d^3\vec{r} = \int \int \int dx\,dy\,dz$

$$= \int_{+\infty}^{-\infty} (-dx')(-dy')(-dz') = \int d^3\vec{r}'$$

| since V' is periodic, the inversion does not matter

$$\int \phi^*(|\vec{r}|)V'\phi(|\vec{r} - a\hat{x}|)d^3\vec{r} = \int \phi^*(|\vec{r}'|)V'\phi(|\vec{r}' + a\hat{x}|)d^3\vec{r}'$$

| relabel $\vec{r}'$ as $\vec{r}$

$$\int \phi^*(|\vec{r}|)V'\phi(|\vec{r} - a\hat{x}|)d^3\vec{r} = \int \phi^*(|\vec{r}|)V'\phi(|\vec{r} + a\hat{x}|)d^3\vec{r}$$

Remarks:
- We do not see a bandgap in this expression because this derivation involves only one type of atomic state and so this $\varepsilon_{\vec{k}}$ is just for one band. Example: s-atomic states form the s-band.

- The transfer integrals directly affect the band width (max $\varepsilon_{\vec{k}}$ - min $\varepsilon_{\vec{k}}$) because the transfer integrals determine the amplitudes of the cosine function.
- The effective mass matrix (to be discussed later in section 3.2.3.1) is calculated (in an exercise) to be

$$m^* = \begin{pmatrix} \dfrac{\hbar^2}{2t_x a^2 \cos(k_x a)} & 0 & 0 \\ 0 & \dfrac{\hbar^2}{2t_y b^2 \cos(k_y b)} & 0 \\ 0 & 0 & \dfrac{\hbar^2}{2t_z c^2 \cos(k_z c)} \end{pmatrix}. \tag{3.166}$$

- Comparison with the nearly free electron model:
  - Both dispersion relations have minima and maxima at the BZ centres and the BZ boundaries, respectively.
  - The maxima and minima are approximately parabolic (to be shown in an exercise).
  - The nearly free electron model and the tight-binding model are band models of opposite extremes.

3.2.2.3 The Kronig–Penney model

To convince you that the physics of the formation of bands and bandgaps is real, we shall solve an exact problem. This is the Kronig–Penney model, which is an electron in a 1D array of periodic rectangular barriers (figure 3.17).

We start by writing down the (1D time-independent) Schrödinger equation for each region.

$$\text{In region (I):} \quad -\frac{\hbar^2}{2m}\frac{d^2}{dx^2}\psi_I(x) = E\psi_I(x) \tag{3.167}$$

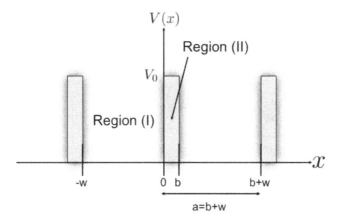

Figure 3.17. The setup for the Kronig–Penney model.

$$\frac{d^2}{dx^2}\psi_I(x) = -q^2\psi_I(x) \quad \text{where } q^2 = \frac{2mE}{\hbar^2} \tag{3.168}$$

$$\psi_I(x) = Ae^{iqx} + Be^{-iqx} \tag{3.169}$$

In region (II): $\quad -\frac{\hbar^2}{2m}\frac{d^2}{dx^2}\psi_{II}(x) + V_0\psi_{II}(x) = E\psi_{II}(x) \quad \text{where } 0 < E < V_0$

$$\frac{d^2}{dx^2}\psi_{II}(x) = Q^2\psi_{II}(x) \quad \text{where } Q^2 = \frac{2m(V_0 - E)}{\hbar^2} \tag{3.170}$$

$$\psi_{II}(x) = Ce^{Qx} + De^{-Qx} \tag{3.171}$$

The four arbitrary constants A, B, C, and D are chosen to fit the four boundary conditions:

$$\text{Wavefunction matching: } \psi_I(0) = \psi_{II}(0) \tag{3.172}$$

$$\text{Derivative of wavefunction matching: } \frac{d\psi_I}{dx}\bigg|_{x=0} = \frac{d\psi_{II}}{dx}\bigg|_{x=0} \tag{3.173}$$

$$\text{Bloch's theorem for the wavefunction: } \psi_{II}(b) = e^{ika}\psi_I(-w) \tag{3.174}$$

$$\text{Bloch's theorem for the derivative of the wavefunction: } \frac{d\psi_{II}}{dx}\bigg|_{x=b} = e^{ika}\frac{d\psi_I}{dx}\bigg|_{x=-w} \tag{3.175}$$

Skipping some algebraic details here, you should get these four equations:

$$A + B = C + D \tag{3.176}$$

$$Aiq - Biq = CQ - DQ \tag{3.177}$$

$$Ce^{Qb} + De^{-Qb} = e^{ika}(Ae^{-iqw} + Be^{iqw}) \tag{3.178}$$

$$CQe^{Qb} - DQe^{-Qb} = e^{ika}(Aiqe^{-iqw} - Biqe^{iqw}) \tag{3.179}$$

A solution for E (hiding inside q and Q) exists when the determinant of the coefficients is zero.

$$\det \begin{vmatrix} 1 & 1 & -1 & -1 \\ iq & -iq & -Q & Q \\ e^{ika-iqw} & e^{ika+iqw} & -e^{Qb} & -e^{-Qb} \\ iqe^{ika-iqw} & -iqe^{ika+iqw} & -Qe^{Qb} & Qe^{-Qb} \end{vmatrix} = 0 \tag{3.180}$$

Evaluating the determinant gives

$$\frac{Q^2 - q^2}{2qQ} \sinh(Qb)\sin(qw) + \cosh(Qb)\cos(qw) = \cos(ka). \qquad (3.181)$$

This expression can give us the dispersion relation E versus k if we solve it numerically. Recall that E is hiding in q and Q. However, this numerical work is quite complicated.

We shall turn this into another exact solution for discussion. We turn the rectangular barriers into Dirac delta barriers with $b \rightarrow 0$ (so $w \rightarrow a$) with the constraint that the area $V_0 b$ is constant. We get

$$\frac{mV_0 ba}{\hbar^2} \frac{\sin(qa)}{qa} + \cos(qa) = \cos(ka). \qquad (3.182)$$

To see the bandgap, refer to figure 3.18. The RHS $=\cos(ka)$ and can thus range between ± 1, while the LHS $=\frac{mV_0 ba}{\hbar^2}\frac{\sin(qa)}{qa} + \cos(qa)$ can range beyond ± 1. The intersections represent solutions for E; however, there are k values (as k is continuous) with no intersections and thus no solutions for E. These are the energy bandgaps.

Thus, this exact model should make it clear that the creation of the energy gap or bandgap is a real physical effect.

Summary: We have described three models for calculating electron energy levels in a periodic potential. We must realise that the two approximations, namely neglecting V_{el-el} and neglecting T_{I-I} (frozen lattice), are too drastic. The modern method that includes V_{el-el} in the calculation of electron energy levels is called density

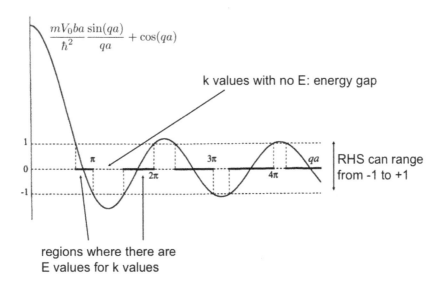

Figure 3.18. The existence of bandgaps in the Kronig–Penney model. The LHS and RHS expressions actually have different ranges, so the allowed energies occur where they have common ranges. Reprinted with permission from [2] John Wiley & Sons.

functional theory (DFT) and because of modern computing power, DFT has become the leading method in determining band structures.

3.2.3 Band theory properties and approximations

Density of states and van Hove singularities

For a solid, the main interest is to calculate the dispersion relation $\varepsilon_{\vec{k}}$. Another important quantity is the density of states $D(\varepsilon)$, as it is also measurable.

Therefore, we now want a general expression that will allow us to calculate the density of states once we know the dispersion relation. We start with the definition of the density of states: the number of states at a certain energy ε (recall equation (2.27)) (figure 3.19).

$$D(\varepsilon) = 2\sum_{\vec{k}}\delta(\varepsilon - \varepsilon_{\vec{k}}),$$

| where two is the spin degeneracy (if present).

| If the solid is infinite, the allowed $\vec{k}$ values are dense.

| Convert the summation to an integral, the volume $d^3\vec{k}$ is compensated by $\dfrac{L_x}{2\pi}\dfrac{L_y}{2\pi}\dfrac{L_z}{2\pi}$. (3.183)

$$D(\varepsilon) = 2\int \frac{L_x}{2\pi}\frac{L_y}{2\pi}\frac{L_z}{2\pi}\delta(\varepsilon - \varepsilon_{\vec{k}})\ d^3\vec{k} \tag{3.184}$$

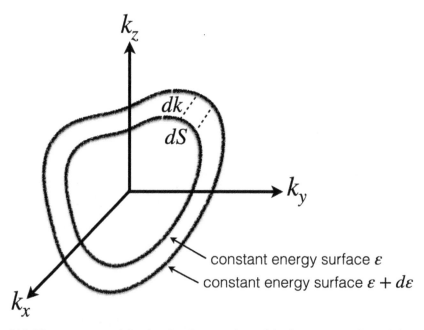

Figure 3.19. The geometry used in changing from an integral in k-space to an integral in E-space.

$$d\varepsilon_{\vec{k}} = \vec{\nabla}_{\vec{k}}\varepsilon_{\vec{k}} \cdot d\vec{k} = \left|\vec{\nabla}_{\vec{k}}\varepsilon_{\vec{k}}\right| dk \implies dk = \frac{d\varepsilon_{\vec{k}}}{\left|\vec{\nabla}_{\vec{k}}\varepsilon_{\vec{k}}\right|}.$$

$$D(\varepsilon) = 2\frac{V}{(2\pi)^3} \int \delta(\varepsilon - \varepsilon_{\vec{k}})\ d^3\vec{k} \tag{3.185}$$

| change into an energy integral (see figure 3.19), where $d^3\vec{k} = dS dk = dS\dfrac{d\varepsilon_{\vec{k}}}{|\vec{\nabla}_{\vec{k}}\varepsilon_{\vec{k}}|}$

$$D(\varepsilon) = 2\frac{V}{(2\pi)^3} \int \frac{\delta(\varepsilon - \varepsilon_{\vec{k}})}{|\vec{\nabla}_{\vec{k}}\varepsilon_{\vec{k}}|}\ dS d\varepsilon_{\vec{k}} \tag{3.186}$$

| carry out the δ −function integral

$$D(\varepsilon) = 2\frac{V}{(2\pi)^3} \int \frac{dS}{|\vec{\nabla}_{\vec{k}}\varepsilon_{\vec{k}}|}\ \Bigg|_{\varepsilon_{\vec{k}}=\varepsilon} \tag{3.187}$$

The quantity $\vec{\nabla}_{\vec{k}}\varepsilon_{\vec{k}}$ is the gradient of the dispersion relation. Singularities in the density of states $D(\varepsilon)$ occur when $\vec{\nabla}_{\vec{k}}\varepsilon_{\vec{k}} = 0$, which is the case when the dispersion has a stationary point[20]. These are called van Hove singularities, and they sometimes show up in optical absorption experiments.[21]

We shall illustrate van Hove singularities for solids of various dimensions.

One-dimensional solid: for a 1D solid, the dispersion relation is really a curve/line and so there are only two possibilities for stationary points where $\dfrac{d\varepsilon_{\vec{k}}}{dk} = 0$: the minimum point at E_0 (which we will call M_0) and the maximum point at E_1 (which we will call M_1).

We assume that near E_0 and E_1, we may approximate the dispersion by parabolas,[22] so that near E_0,

$$\varepsilon_{\vec{k}} \approx E_0 + \frac{\hbar^2 k_x^2}{2m}, \quad \text{which is, of course, at a minimum at } E^0. \tag{3.188}$$

So, we work out the DOS near E_0:

$$D(\varepsilon) = 2\sum_{\vec{k}} \delta(\varepsilon - \varepsilon_{\vec{k}}) \tag{3.189}$$

| assume k is dense and convert it into integral

$$D(\varepsilon) = 2\frac{L_x}{2\pi} \int \delta\left(\varepsilon - E_0 - \frac{\hbar^2 k_x^2}{2m}\right) dk_x \tag{3.190}$$

| convert to energy integral, let $\varepsilon' = \dfrac{\hbar^2 k_x^2}{2m}$ and $dk_x = \sqrt{\dfrac{2m}{\hbar^2}}\dfrac{1}{2\sqrt{\varepsilon'}}d\varepsilon'$

| need to include a factor of two for both positive and negative k_x

---

[20] The expression $\vec{\nabla}_{\vec{k}}\varepsilon_{\vec{k}}$ is also the group velocity (section 3.2.3.1), so van Hove singularities occur when the group velocity is zero.
[21] This is because the stationary points $\vec{\nabla}_{\vec{k}}\varepsilon_{\vec{k}} = 0$ are typically at the topmost part of the valence band and the bottommost part of the conduction band. This is the region where optical transitions usually occur.
[22] See 'Remarks' at the end of the tight-binding model.

$$D(\varepsilon) = 2\frac{L_x}{2\pi}\sqrt{\frac{2m}{\hbar^2}} \int_0^\infty \delta(\varepsilon - E_0 - \varepsilon')\frac{1}{\sqrt{\varepsilon'}} \ d\varepsilon' \tag{3.191}$$

$$D(\varepsilon) = 2\frac{L_x}{2\pi}\sqrt{\frac{2m}{\hbar^2}}\frac{1}{\sqrt{\varepsilon - E_0}} \quad \text{with } \varepsilon > E_0 \tag{3.192}$$

Then for E_1, $\varepsilon_{\vec{k}} \approx E_1 - \frac{\hbar^2 k_x^2}{2m}$, and the DOS is similarly determined (figure 3.20),

$$D(\varepsilon) = 2\frac{L_x}{2\pi} \int \delta\left(\varepsilon - E_1 + \frac{\hbar^2 k_x^2}{2m}\right)dk_x \tag{3.193}$$

$$D(\varepsilon) = 2\frac{L_x}{2\pi}\sqrt{\frac{2m}{\hbar^2}}\frac{1}{\sqrt{E_1 - \varepsilon}} \quad \text{with } \varepsilon < E_1 \tag{3.194}$$

Two-dimensional solid: for a 2D solid, the dispersion relation is made of superposed paraboloids, so we have three types of stationary points: a minimum at E_0 called M_0, a saddle point at E_1 called M_1, and a maximum at E_2 called M_2.

Again, we assume that near E_0, the dispersion is approximately a paraboloid.

$$\varepsilon_{\vec{k}} \approx E_0 + \frac{\hbar^2 k_x^2}{2m} + \frac{\hbar^2 k_y^2}{2m} \tag{3.195}$$

So the DOS near E_0,

$$D(\varepsilon) = 2\sum_{\vec{k}} \delta(\varepsilon - \varepsilon_{\vec{k}}) \tag{3.196}$$

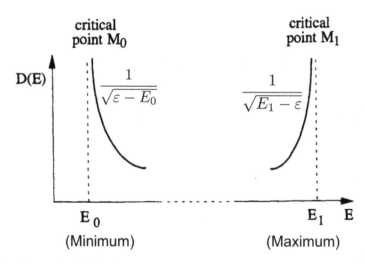

Figure 3.20. Van Hove singularities in 1D. Reprinted from [4], Copyright (2013), with permission from Elsevier.

$$D(\varepsilon) = 2\frac{L_x}{2\pi}\frac{L_y}{2\pi} \int \delta\left(\varepsilon - E_0 - \frac{\hbar^2 k_x^2}{2m} - \frac{\hbar^2 k_y^2}{2m}\right) d^2\vec{k}$$

(3.197)

| change to polar coordinates: $k_x^2 + k_y^2 = k^2$ and $d^2\vec{k} = k d\theta dk$

| integrate over θ to get 2π

$$D(\varepsilon) = 2\frac{L_x L_y}{2\pi} \int_0^\infty \delta\left(\varepsilon - E_0 - \frac{\hbar^2 k^2}{2m}\right) k\ dk$$

(3.198)

|convert to energy integral: $\varepsilon' = \frac{\hbar^2 k^2}{2m}$ and $dk = \sqrt{\frac{2m}{\hbar^2}}\frac{1}{2\sqrt{\varepsilon'}}d\varepsilon'$

$$D(\varepsilon) = 2\frac{L_x L_y}{2\pi}\sqrt{\frac{2m}{\hbar^2}}\sqrt{\frac{2m}{\hbar^2}} \int_0^\infty \delta(\varepsilon - E_0 - \varepsilon')\frac{1}{2\sqrt{\varepsilon'}}\sqrt{\varepsilon'}\ d\varepsilon'$$

(3.199)

$$D(\varepsilon) = \frac{L_x L_y}{2\pi}\frac{2m}{\hbar^2}$$

(3.200)

which is a constant type of 'singularity'.

For the maximum M_2 at E_2, we have

$$\varepsilon_{\vec{k}} \approx E_0 - \frac{\hbar^2 k_x^2}{2m} - \frac{\hbar^2 k_y^2}{2m}$$

(3.201)

$$D(\varepsilon) = 2\frac{L_x}{2\pi}\frac{L_y}{2\pi} \int \delta\left(\varepsilon - E_0 + \frac{\hbar^2 k_x^2}{2m} + \frac{\hbar^2 k_y^2}{2m}\right) d^2\vec{k}$$

(3.202)

| we are going to end up with the same constant

$$D(\varepsilon) = \frac{L_x L_y}{2\pi}\frac{2m}{\hbar^2}$$

(3.203)

For the saddle point M_1 at E_1, we have[23]

$$\varepsilon_{\vec{k}} \approx E_0 + \frac{\hbar^2 k_x^2}{2m} - \frac{\hbar^2 k_y^2}{2m}$$

(3.204)

---

[23] A saddle point is essentially the combination of a maximum point in one direction and a minimum point in another direction.

$$D(\varepsilon) = 2\frac{L_x}{2\pi}\frac{L_y}{2\pi}\int \delta\left(\varepsilon - E_0 - \frac{\hbar^2 k_x^2}{2m} + \frac{\hbar^2 k_y^2}{2m}\right)d^2\vec{k}$$

| work in Cartesian coordinates: $d^2\vec{k} = dk_x dk_y$

| let $\varepsilon_x = \frac{\hbar^2 k_x^2}{2m}$, $\varepsilon_y = \frac{\hbar^2 k_y^2}{2m}$ so $dk_x = \sqrt{\frac{2m}{\hbar^2}}\frac{1}{2\sqrt{\varepsilon_x}}d\varepsilon_x$ (3.205)

| and $dk_y = \sqrt{\frac{2m}{\hbar^2}}\frac{1}{2\sqrt{\varepsilon_y}}d\varepsilon_y$

$$D(\varepsilon) = 2\frac{L_x}{2\pi}\frac{L_y}{2\pi}\int \delta(\varepsilon - E_1 - \varepsilon_x + \varepsilon_y)\frac{2m}{\hbar^2}\frac{1}{2\sqrt{\varepsilon_x}}\frac{1}{2\sqrt{\varepsilon_y}}d\varepsilon_x d\varepsilon_y$$ (3.206)

| carry out the ε_x integration using the delta function

$$D(\varepsilon) = m\frac{L_x L_y}{\pi^2 \hbar^2}\int \frac{1}{\sqrt{\varepsilon - E_1 + \varepsilon_y}}\frac{1}{\sqrt{\varepsilon_y}}d\varepsilon_y$$

| complete the square: $\varepsilon_y(\varepsilon_y + \varepsilon - E_1) = \left(\varepsilon_y + \frac{\varepsilon - E_1}{2}\right)^2 - \left(\frac{\varepsilon - E_1}{2}\right)^2$ (3.207)

| then use the integral identity: $\int \frac{1}{\sqrt{x^2 \pm a^2}}dx = \ln\left|x + \sqrt{x^2 \pm a^2}\right|$

$$D(\varepsilon) = m\frac{L_x L_y}{\pi^2 \hbar^2}\int_0^\alpha \frac{1}{\sqrt{\left(\varepsilon_y + \frac{\varepsilon - E_1}{2}\right)^2 - \left(\frac{\varepsilon - E_1}{2}\right)^2}}d\varepsilon_y$$

(3.208)

| the upper limit will blow up the integral, so we let it be α first

| change variables to $\varepsilon' = \varepsilon_y + \frac{\varepsilon - E_1}{2}$ then $\int_0^\alpha \longrightarrow \int_{(\varepsilon - E_1)/2}^{\alpha' = \alpha + (\varepsilon - E_1)/2}$

$$D(\varepsilon) = m\frac{L_x L_y}{\pi^2 \hbar^2}\int_{(\varepsilon - E_1)/2}^{\alpha'} \frac{1}{\sqrt{\varepsilon_y'^2 - \left(\frac{\varepsilon - E_1}{2}\right)^2}}d\varepsilon_y'$$ (3.209)

$$D(\varepsilon) = m\frac{L_x L_y}{\pi^2 \hbar^2}\left[\ln\left|\varepsilon_y' + \sqrt{\varepsilon_y'^2 - \left(\frac{\varepsilon - E_1}{2}\right)^2}\right|\right]_{(\varepsilon - E_1)/2}^{\alpha'}$$ (3.210)

$$D(\varepsilon) = m\frac{L_x L_y}{\pi^2 \hbar^2}\left(\ln\left|\alpha' + \sqrt{\alpha'^2 - \left(\frac{\varepsilon - E_1}{2}\right)^2}\right| - \ln\left|\frac{\varepsilon - E_1}{2}\right|\right)$$ (3.211)

| take it near $\varepsilon - E_1 \approx 0$

$$D(\varepsilon) = m\frac{L_x L_y}{\pi^2 \hbar^2} \ln \frac{4\alpha'}{|\varepsilon - E_1|} \tag{3.212}$$

which is a logarithmic divergence. The diagrams in figure 3.21 illustrate the van Hove singularities in 2D.

Three-dimensional solid: for 3D, as the calculations are tedious but similar, we only state the schematics here. It should be easy to guess that there are four types of critical points: a minimum at E_0 called M_0, two saddle points (one at E_1, called M_1, and one at E_2, called M_2), and a minimum at E_3 called M_3. The calculation is essentially the calculation of the DOS of the free electron gas, so there should be a $\sqrt{\varepsilon}$ expression (recall equation (3.65)). Diagrams of the van Hove singularities in 3D are shown in figure 3.22. A real band diagram with its DOS is shown in figure 3.23.

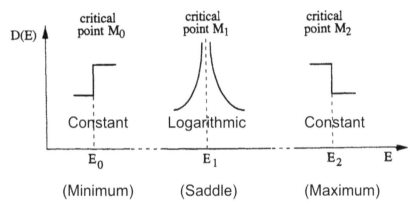

Figure 3.21. Van Hove singularities in 2D. Reprinted from [4], Copyright (2013), with permission from Elsevier.

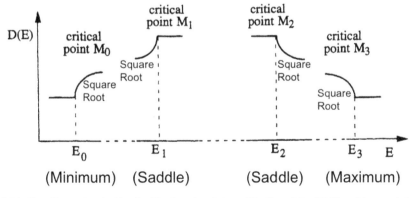

Figure 3.22. Van Hove singularities in 3D. Reprinted from [4], Copyright (2013), with permission from Elsevier.

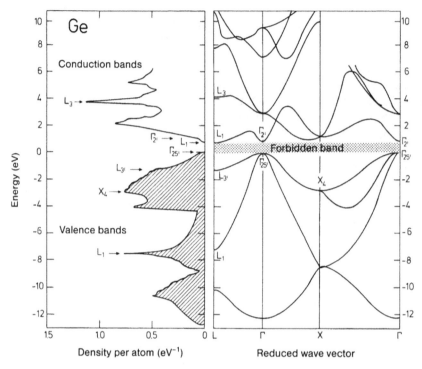

Figure 3.23. Here is a real band diagram (RIGHT) with the DOS (LEFT). The labels on the DOS diagram are the van Hove singularities due to the specific stationary points shown on the band diagram. Reprinted by permission from Springer Nature [5], Copyright (2009).

3.2.3.1 The effective mass approximation

We recall $\psi(\vec{r}) = \sum_{\vec{k}} C_{\vec{k}} e^{i \vec{k} \cdot \vec{r}}$ as the form of the Bloch function that we started out using. It is a superposition of plane waves. Whenever we have a superposition of plane waves, it is useful to define the group velocity $\vec{v}_g$.[24]

In this case for the electron in a periodic potential (a Bloch electron), the group velocity is defined as $\vec{v}_g = \frac{1}{\hbar} \vec{\nabla}_{\vec{k}} \varepsilon_{\vec{k}}$, which is essentially proportional to the gradient of the dispersion relation.

Now that we have a notion of the 'velocity of an electron in a periodic potential', we shall go further and work out the (semiclassical) dynamics of the Bloch electron. We shall make a reasonable assumption that the electron does not make interband transitions; in other words, it stays in one band.

We define the acceleration (of an electron in a periodic potential) in an intuitive way:

---

[24] Recall that in section 2.1.2, the phonon group velocity was defined as $v_g = \frac{d\omega}{dk}$. Using $E = \hbar\omega$ and generalising to 3D yields the expression $\vec{v}_g = \frac{1}{\hbar} \vec{\nabla}_{\vec{k}} E$.

$$\vec{a} = \frac{d\vec{v}_g}{dt} = \frac{1}{\hbar}\frac{d}{dt}\vec{\nabla}_{\vec{k}}\varepsilon_{\vec{k}}$$

$\mid$ use thechain rule: $\dfrac{d}{dt} = \dfrac{d\vec{k}}{dt}\cdot\vec{\nabla}_{\vec{k}} = \sum_{i=1}^{3}\dfrac{dk_i}{dt}\dfrac{\partial}{\partial k_i}$ (3.213)

$\mid$ and $\vec{\nabla}_{\vec{k}}\varepsilon_{\vec{k}} = \sum_{j=1}^{3}\dfrac{\partial\varepsilon_{\vec{k}}}{\partial k_j}\hat{e}_j = \dfrac{\partial\varepsilon_{\vec{k}}}{\partial k_x}\hat{x} + \dfrac{\partial\varepsilon_{\vec{k}}}{\partial k_y}\hat{y} + \dfrac{\partial\varepsilon_{\vec{k}}}{\partial k_z}\hat{z}$

$$\vec{a} = \frac{1}{\hbar}\sum_i \frac{dk_i}{dt}\frac{\partial}{\partial k_i}\left(\sum_{j=1}^{3}\frac{\partial\varepsilon_{\vec{k}}}{\partial k_j}\hat{e}_j\right)$$

$\mid$ note the force expression $\hbar\dfrac{dk_i}{dt} = F_i$ (3.214)

$\mid$ based on Newton's second law,

$\mid$ define an inverse effective mass matrix as $(m^{*-1})_{ij} = \dfrac{1}{\hbar^2}\dfrac{\partial^2\varepsilon_{\vec{k}}}{\partial k_i\partial k_j}$

$$\vec{a} = \sum_{i,j}F_i(m^{*-1})_{ij}\hat{e}_j$$ (3.215)

The whole point is that now we have some sort of Newton's second law for the electron in the periodic potential. The effects of the periodic potential are captured in the effective mass matrix. This enables us to perform a simple analysis of the electron in the solid.

Remarks:

- If the dispersion is isotropic, meaning $\varepsilon_{\vec{k}}$ only depends on $|\vec{k}|$, then the effective mass matrix m^* is diagonal, since $\dfrac{\partial^2\varepsilon_{\vec{k}}}{\partial k_i\partial k_j} = 0$ for $i \neq j$. Also $\dfrac{\partial^2\varepsilon_{\vec{k}}}{\partial k_i^2}$ are the same for $i = x, y, z$, so the matrix m^* is a number $\times$ identity matrix. Thus, m^* can be written as $m^* = \dfrac{\hbar^2}{\frac{d^2\varepsilon_{\vec{k}}}{dk^2}}$.

- Recall that the dispersion relation is parabolic near the maximum or the minimum of a band, i.e.

$$\varepsilon_{\vec{k}} \approx E_0 + \frac{\hbar^2}{2m}(\vec{k} - \vec{k}_0)^2$$ (3.216)

at the wave vector $\vec{k}_0$. It would be a better approximation to write $m \to m^*$ so as to include the effects of the periodic potential, as follows:

$$\varepsilon_{\vec{k}} \approx E_0 + \frac{\hbar^2}{2m^*}(\vec{k} - \vec{k}_0)^2$$ (3.217)

- Similarly, the DOS (near the maximum or minimum of a band) can be approximated by the free electron DOS if m is replaced by m^* so as to include the effects of the periodic potential.

$$D(\varepsilon) = \frac{V}{2\pi^2}\left(\frac{2m^*}{\hbar^2}\right)^{3/2}\sqrt{\varepsilon} \qquad (3.218)$$

You can think of this as follows (approximately speaking): an electron is in a periodic potential and so it experiences a 'change in mass'. The effect of the periodic potential on the free electron DOS is simply to modify the mass parameter.

3.2.3.2 The concept of a hole

If we have a band that is mostly filled and only has a few empty states close to the top, then it is useful to introduce the concept of a hole rather than talking about electrons, since most of them cannot participate in physical processes (figure 3.23).

Consider a parabolic band with its maximum at the origin of the E–$\vec{k}$ axes (figure 3.24).

The wave vector of a hole: If this band is full (of electrons), the sum over $\vec{k}$ is zero: $\sum_j \vec{k}_j = 0$.

Suppose an electron with $\vec{k}_l$ is removed; then, the net $\vec{k}$ of the band is

$$\text{net } \vec{k} = \sum_{j(\neq l)} \vec{k}_j = \sum_j \vec{k}_j - \vec{k}_l = 0 - \vec{k}_l = -\vec{k}_l = \vec{k}_h, \qquad (3.219)$$

where we define the wave vector of the hole $\vec{k}_h$ as the net wave vector of the band.

The energy of a hole: The hole's energy can be defined as the energy required to flush out the electron $\vec{k}_l$, so $\varepsilon_h = -\varepsilon_{\vec{k}_l}$. Note that in our setup (figure 3.24), $\varepsilon_{\vec{k}_l} < 0$, so that in reality, $\varepsilon_h > 0$.

The group velocity of a hole: $\vec{v}_h = \frac{1}{\hbar}\vec{\nabla}_{\vec{k}_h}\varepsilon_h = \frac{1}{\hbar}\vec{\nabla}_{-\vec{k}_l}(-\varepsilon_{\vec{k}_l}) = \vec{v}_l$ where the two negative signs cancel out. The group velocity of the hole $\vec{v}_h$ is thus the same as the group velocity of the missing electron.

Current: the full band carries no current:

$$\text{net current} = \sum_j (-e)\vec{v}_j = 0 \qquad (3.220)$$

After the removal of the electron $\vec{k}_l$:

$$\text{net current} = \sum_{j(\neq l)} (-e)\vec{v}_j = -(-e)\vec{v}_l = +e\vec{v}_h \qquad (3.221)$$

Thus, the hole appears to be associated with a positive charge.

The effective mass of a hole: The effective mass of the hole is the negative of the effective mass of the missing electron $\vec{k}_l$ (exercise):

$$m_h^* = -m_l^* \qquad (3.222)$$

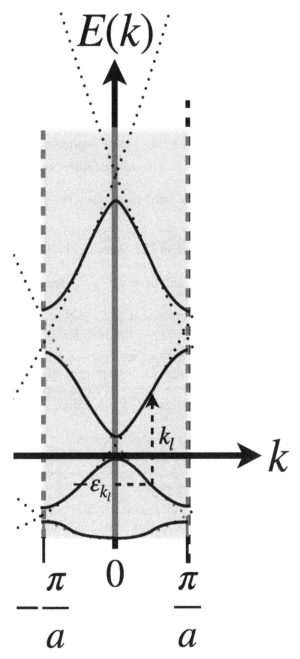

Figure 3.24. A schematic that illustrates a hole as a missing electron with wave vector $\vec{k}_l$ in an almost filled band.

It is interesting to note that the concept of holes actually explains the fact that many divalent and trivalent metals have positive Hall coefficients. A simple explanation is as follows:

- The Fermi spheres of these metals are large enough to 'overflow' into the 2BZ (figure 3.15). The 'overflowing' electrons contribute to the negative Hall coefficient.
- The corners of the 1BZ are unfilled electron states or hole states that contribute to a positive Hall coefficient.
- Their relative contributions decide the net Hall coefficient.

3.2.4 A brief mention of the measurement of band structures

For completeness, I would just like to briefly mention how band structures are measured experimentally.

Optical absorption: this method is useful for probing the bandgap of an insulator or semiconductor.

- Light is incident on the solid, and no light is absorbed if the energy of the photon ($E = hf$) is smaller than the bandgap. This photon simply passes through the solid.
- Absorption becomes significant once hf equals the bandgap. Hence, the bandgap can be determined.
- The relationship between absorption and the thickness of the solid x can be described by $\frac{I(x)}{I_0} = e^{-\alpha x}$, where $I(x)$ is the outgoing intensity after the photon has passed through the solid, I_0 is the incident intensity, and α is the so-called absorption coefficient. Measuring the absorption coefficient (for incident light of various frequencies) also reveals the van Hove singularities (section 3.2.3) because a sudden change in DOS changes the absorption behaviour.

Cyclotron resonance: this is the older standard method.

- A magnetic field is applied and the (free) electron rotates in a circular orbit, perpendicular to the magnetic field, with the cyclotron frequency $\omega_c = \frac{eB}{m}$ (recall right below equation (3.22)).
- This cyclotron frequency is measured by how well the sample absorbs microwave frequencies. The cyclotron resonance method thus probes the (constant) Fermi energy surface.
- For semiconductors, the effective mass can be measured as well.
- Cyclotron resonance is not further discussed, as it would require an understanding of the quantum effect of the magnetic field on the energy levels. The energy levels become so-called Landau levels. For further details, the student may refer to a more advanced textbook, such as [6].

Angle-resolved photoemission spectroscopy (ARPES): This is the modern standard method, as it can probe the dispersion relation in great detail.

- This method effectively makes use of photoelectric effect. A photon is absorbed by the (valence) electron, and the electron leaves the solid. The initial state of the electron is a Bloch state and the final state is a free electron.
- The reasons for its great usefulness are:

- The photon's energy $hf = \hbar\omega$ and wave vector $\vec{k}_{photon}$ are known. Actually, $\vec{k}_{photon}$ is very small.
    - The electron that leaves the solid has its angle of emission and kinetic energy measured.
    - The electron wave refracts at the surface; for a clean surface, the component of the wave vector parallel to the surface is conserved. Then conservation of energy is used as well.
    - With all this information, it is possible to map out the band structure.
- Typically, a synchrotron light source is used because it is intense and monochromatic and, most importantly, its frequency can be varied continuously.

3.3 Exercises

3.3.1 Question: the 3D Sommerfeld model

(Solution in section 6.3.1.) For the 3D quantum free electron gas, show that the internal energy at $T = 0$ can be written as $U(0) = \frac{3}{5}N\varepsilon_F$.

3.3.2 Question: the 2D Sommerfeld model

(Solution in section 6.3.2.) For a 2D quantum free electron gas confined in a box $L_x \times L_y$:

1. If there are N electrons and the electron density is defined as $n = \frac{N}{L_x L_y}$, show that the Fermi wave vector is $k_F = (2\pi n)^{1/2}$.
2. Show that the DOS is a constant $D(\varepsilon) = \frac{mL_x L_y}{\pi \hbar^2}$.
3. It turns out that the constant DOS in 2D allows the chemical potential to be solved exactly as a function of temperature. Show that the 2D chemical potential is exactly $\mu = k_B T \ln\left(\exp\left(\frac{N\pi\hbar^2}{L_x L_y m k_B T}\right) - 1\right)$.

3.3.3 Question: sketching 2D bands in the empty lattice approximation

(Solution in section 6.3.3.) For a 2D square lattice (in real space) of spacing a, sketch the electron bands in the empty lattice approximation for the cross section $\Gamma - M - X - \Gamma$. Sketch up to $E \approx 10\frac{h^2}{8ma^2}$. Hint: You will need to consider dispersion curves from several neighbouring BZs.

3.3.4 Question: the nearly free electron model in 2D

(Solution in section 6.3.4.) Consider a 2D square lattice of side a.

1. In the empty lattice approximation, show that the free electron's energy at the M point of the 1BZ is twice that at the X point of the 1BZ.
2. Assume the crystal potential of this solid is

$$V(x, y) = 2V_0\left(\cos\frac{2\pi x}{a} + \cos\frac{2\pi y}{a}\right), \tag{3.223}$$

where V_0 is a constant. Write down the bandgap at the X point with the lowest energy.

3. As discussed for figure 3.15, there is band overlap that results in Group 2 elements being metals. Which parameter (in the above potential), if tuned, can turn these 2D solids into insulators? Do we tune it up or down?

3.3.5 Question: the 1D tight-binding model

(Solution in section 6.3.5.) Assume a 1D solid of spacing a.
1. Write down the dispersion relation and sketch its 1BZ.
2. Give an expression for the width of the band (i.e. the highest energy in the band minus the lowest energy in the band). This is also sometimes called the band width.
3. Sketch the group velocity of the electron in the 1BZ.

3.3.6 Question: comparing the tight-binding and nearly free electron models

(Solution in section 6.3.6.) The basic similarity between these models is that for both models, the dispersion can be approximated by parabolas ($E \propto k^2$) at the minimum (at the BZ centre) and at the maximum (at the BZ edge). Let us check this using a 1D solid.
1. The nearly free electron model: at the BZ centre, this model is a parabola, so there is nothing to check. Near the BZ edge, recall the dispersion relation (equation (3.146)):

$$E = \frac{1}{2}\left(E_{\vec{q}-\vec{G}'_a}^{\text{free}} + E_{\vec{q}-\vec{G}'_b}^{\text{free}}\right) \pm \sqrt{\left(\frac{E_{\vec{q}-\vec{G}'_a}^{\text{free}} - E_{\vec{q}-\vec{G}'_b}^{\text{free}}}{2}\right)^2 + |V_{\vec{G}'_a-\vec{G}'_b}|^2} \tag{3.224}$$

and show that near $q = \frac{\pi}{a}$, you get an approximate parabola.
2. The tight-binding model: at the BZ centre and the BZ edge, show that the 1D tight-binding dispersion can be approximated by parabolas.

3.3.7 Question: the Kronig–Penney model

(Solution in section 6.3.7.) Fill in the steps that are missing in the discussion of the Kronig–Penney Model on section 3.2.2.3.
1. Derive the four equations from the four boundary conditions.
2. Work out the determinant of the coefficients by hand and arrive at

$$\frac{Q^2 - q^2}{2qQ}\sinh(Qb)\sin(qw) + \cosh(Qb)\cos(qw) = \cos(ka). \tag{3.225}$$

3. Show how the above expression becomes

$$\frac{mV_0ba}{\hbar^2}\frac{\sin(qa)}{qa} + \cos(qa) = \cos(ka) \tag{3.226}$$

in the limit of Dirac delta function barriers.

3.3.8 Question: the tight-binding effective mass

(Solution in section 6.3.8.)

1. Derive the effective mass matrix (equation (3.166)) and show that it is as described in the discussion of the tight-binding model:

$$m^* = \begin{pmatrix} \dfrac{\hbar^2}{2t_xa^2\cos(k_xa)} & 0 & 0 \\[2ex] 0 & \dfrac{\hbar^2}{2t_yb^2\cos(k_yb)} & 0 \\[2ex] 0 & 0 & \dfrac{\hbar^2}{2t_zc^2\cos(k_zc)} \end{pmatrix}.$$

2. Assuming the lattice spacings are all equal ($a = b = c$) and the hopping integrals are all equal ($t_x = t_y = t_z = t$), write down the effective mass.

References

[1] Alloul H 2010 *Introduction to the Physics of Electrons in Solids* (Berlin: Springer)
[2] Kittel C 2005 *Introduction to Solid State Physics* 8th edn (New York: Wiley) https://www.wiley.com/en-ie/Kittel's+Introduction+to+Solid+State+Physics,+8th+Edition,+Global+Edition-p-9781119454168
[3] Ziman J M 1972 *Principles of the Theory of Solids* (Cambridge: Cambridge University Press)
[4] Grosso G and Parravicini G P 2013 *Solid State Physics* 2nd edn (New York: Academic)
[5] Ibach H and Lüth H 2009 *Solid-State Physics: An Introduction to Principles of Materials Science* (Berlin: Springer)
[6] Ashcroft N W and Mermin N D 2022 *Solid State Physics* (London: Cengage Learning)

IOP Publishing

Crystalline Solid State Physics
An interactive guide
Meng Lee Leek

Chapter 4

Basic semiconductor physics

4.1 Homogeneous semiconductors

Semiconductors are important in driving our technology because their electronic properties can be manipulated to our needs. Hence, it is important to understand the electronic properties of semiconductors so that we can manipulate these properties.

'Homogeneous' semiconductors refers to semiconductors whose type is the same across the entire piece of material. Here, the same type can mean either the same pure material (an intrinsic semiconductor) or same doped material (an extrinsic semiconductor). We shall consider intrinsic semiconductors first.

4.1.1 Intrinsic semiconductors

4.1.1.1 A discussion of the standard examples: silicon and germanium
In terms of technological applications, the most important examples of semiconductors are silicon (Si), germanium (Ge), and gallium arsenide (GaAs) because they are widely used in electronics today (figure 4.1).

These materials have even numbers of valence electrons per primitive cell; therefore, their valence bands are completely filled (at $T = 0$) and electrons need to jump across the bandgap to reach the conduction band. The most important parts of the dispersion for technological purposes are the top of valence band (VB) and the bottom of the conduction band (CB), simply because this region has the smallest energy requirement for the electron to jump between bands. One important complication is that the top of VB and the bottom of CB may be at different k-points. This is called an indirect bandgap. Silicon and germanium are the standard examples of indirect bandgap semiconductors.

When the top of the VB and the bottom of the CB are at roughly the same k-point, we call this a direct bandgap. GaAs is the standard example of a direct bandgap semiconductor.

We shall now discuss the basic features of direct and indirect semiconductors.

doi:10.1088/978-0-7503-5217-8ch4
© IOP Publishing Ltd 2023

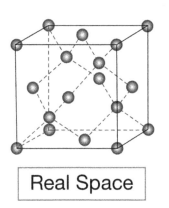
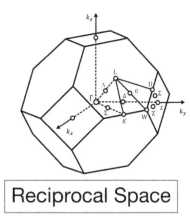

Real Space | **Reciprocal Space**

Figure 4.1. The real spaces of Si, Ge, and GaAs are FCC with a two-atom basis, so their 1BZs are BCC.

4.1.1.2 The constant energy surface

- At $T = 0$, the valence bands are completely filled with electrons and the conduction bands are empty.
- As the temperature rises above zero, electrons are thermally excited over the bandgap E_g from the top of the VB to the bottom of the CB. Note that the top of the VB and the bottom of the CB may not be at the same k-point.
- The population of electrons at the bottom of CB forms the 'Fermi surface' or constant energy surface of the semiconductor, as these are the electrons that participate in the semiconducting properties.[1]
- Note that the bottom of the CB can be located at several k-points related by symmetries. Thus, the constant energy surface repeats at those symmetry-related k-points (figures 4.2, 4.3, and 4.4).

4.1.1.3 The use of optical absorption to probe the bandgap and band structure

- An electron at the top of VB absorbs a photon and is excited to the bottom of the CB. This transition requires the smallest amount of energy.
- Note that the wave vector k of a photon is about 1000 times smaller than the width of the BZ. So the optical transition results in almost no change in the wave vector k of the electron. We call an optical transition a 'vertical transition'.
- The k-point of the top of the VB and the k-point of the bottom of the CB may or may not be the same, so we have two types of transition from the VB to the CB.

    Direct transition: this occurs when the k-point at the top of the VB and the k-point at the bottom of the CB are about the same. Hence, absorbing a photon alone fulfils both energy and wave vector conservation (figure 4.5).

---

[1] Be careful of the use of words! In metals, the Fermi surface is the highest constant energy surface. This is why the constant energy surface in semiconductors is sometimes also called the Fermi surface. Actually, it is preferable to reserve the term 'Fermi surface' for metals.

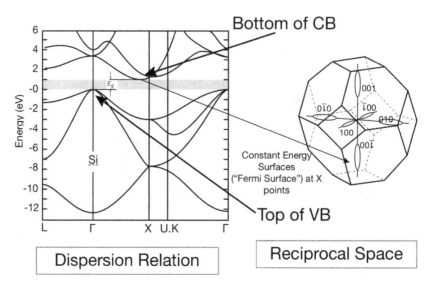

Figure 4.2. The 'Fermi surface' or constant energy surface for Si. Since the top of the CB is almost at the X point and there are six symmetry-related X points, there are therefore six constant energy surfaces. Adapted with permission from [1], Copyright (2009) and [2], Copyright (2004), with permission from Springer Nature.

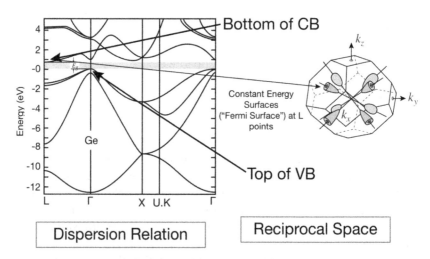

Figure 4.3. The 'Fermi surface' or constant energy surface for Ge. Since the top of the CB is at the L point and there are eight symmetry-related L points, there are therefore eight constant energy surfaces. Adapted with permission from [1], Copyright (2009) and [2], Copyright (2004), with permission from Springer Nature.

Indirect transition: this occurs when the k-point at the top of the VB and the k-point at the bottom of the CB are very different. Hence, absorbing a photon alone cannot fulfil the wave vector conservation condition. The electron also has to absorb a phonon to fulfil wave vector conservation. A phonon can provide a large k but a small energy (figures 4.6 and 4.7).

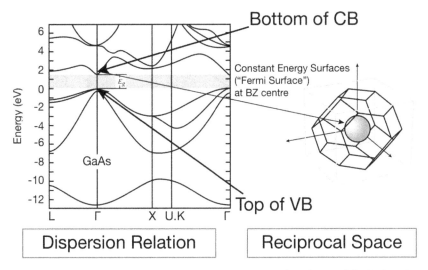

Figure 4.4. The 'Fermi surface' or constant energy surface for GaAs. Since the top of the CB is at the Γ point, and there is only 1 Γ point, there is therefore only one constant energy surface. Reprinted by permission from Springer Nature [1], Copyright (2009).

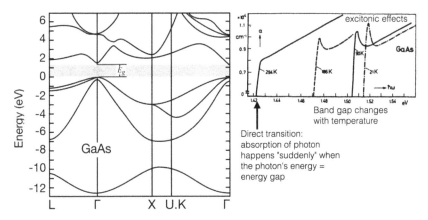

Figure 4.5. A direct transition in GaAs. Reprinted by permission from Springer Nature [2], Copyright (2004).

4.1.1.4 Optical absorption including excitonic effects

- When an electron is excited so that it leaves the VB, there is a hole in VB which is effectively positively charged (see 'The concept of a hole' on page 120).
- If the electron is not energetic enough to be directly excited into the conduction band, the electron and the hole (created in the VB) bind and form a 'positronium atom'. This is called an exciton. See figure 4.8 for basic versions of excitons.

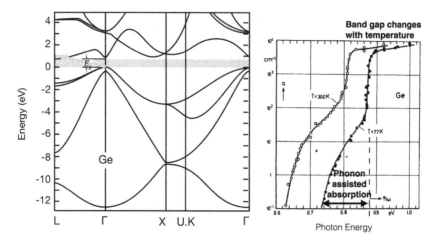

Figure 4.6. An indirect transition in Ge. Optical absorption begins at photon energies less than E_g because the shortfall in energy is supplemented by the phonon. Reprinted by permission from Springer Nature [2], Copyright (2004).

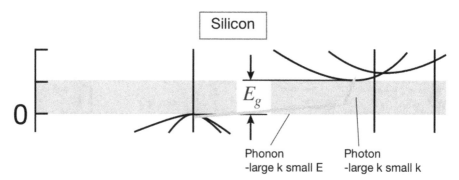

Figure 4.7. The mechanism of indirect transition (using Si as example).

Figure 4.8. Exciton: electron–hole coupling exists in two extreme forms. LEFT: a Wannier exciton, which has a radius of ∼10 nm and a binding energy of ∼10 meV. RIGHT: a Frenkel exciton, which has a radius of ∼1 nm and a binding energy of ∼1 eV.

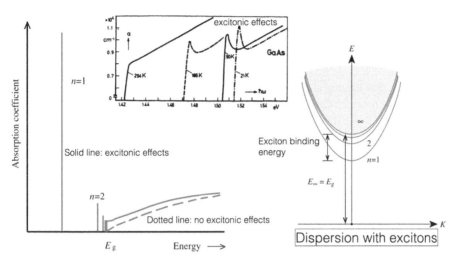

Figure 4.9. Semiconductor optical absorption with excitonic effects. LEFT: absorption spectrum including excitons (solid line). TOP: real absorption spectrum of GaAs showing the appearance of excitons at low temperatures. RIGHT: dispersion relation modified to include exciton energy levels. Reprinted by permission from Springer Nature [3], Copyright (2010).

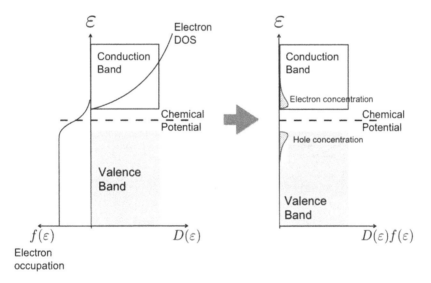

Figure 4.10. A schematic showing the calculation of intrinsic carrier concentration. LEFT: overlay of DOS and occupation probability $f(\varepsilon)$ over the band diagram. RIGHT: the product of the DOS and $f(\varepsilon)$ is related to the concentration of electrons in the CB (and holes in the VB).

- The exciton energy must be a little smaller than the bottom of the CB because it only takes a small thermal energy to break the electron–hole bond. When the exciton is 'broken up', the electron and hole do not bind and they move independently in the CB and the VB, respectively. In figure 4.9, we see that excitons are only observed at low temperatures (as expected).

4.1.1.5 Intrinsic carrier concentration

We now want to calculate the electron carrier concentration in the CB due to thermal excitation and the corresponding hole carrier concentration in the VB.

We assume that the top of VB and bottom of CB dispersions are paraboloids and so the Sommerfeld model is used as an approximation for this calculation. Recall equation (3.79) and noting figure 4.10, the expression used to calculate the total number of particles in thermal equilibrium at temperature T (in the Sommerfeld model):

$$N = \int_{\text{whole range of energy}} D(\varepsilon)f(\varepsilon)d\varepsilon. \tag{4.1}$$

Therefore, for a semiconductor whose bandgap $E_g = E_c - E_v$, we can separately calculate the numbers of carriers in the VB and the CB.

The concentration of electrons in the CB: $n_c = \dfrac{N}{V} = \dfrac{1}{V}\displaystyle\int_{E_c}^{\infty} D(\varepsilon)f(\varepsilon)d\varepsilon$ (4.2)

$$n_c = \frac{1}{V}\int_{E_c}^{\infty} D(\varepsilon)\frac{1}{e^{(\varepsilon-\mu)/k_BT} + 1}d\varepsilon \tag{4.3}$$

The concentration of holes in VB: $p_v = \dfrac{1}{V}\displaystyle\int_{-\infty}^{E_v} D(\varepsilon)(1 - f(\varepsilon))\, d\varepsilon$

(4.4)

| as electron and hole occupancies are opposite

$$p_v = \frac{1}{V}\int_{-\infty}^{E_v} D(\varepsilon)\frac{1}{e^{(\mu-\varepsilon)/k_BT} + 1}d\varepsilon \tag{4.5}$$

We shall calculate n_c. The answer for p_v will be given later and the calculation for p_v shall be done as an exercise.

$$n_c = \frac{1}{V}\int_{E_c}^{\infty} D(\varepsilon)\frac{1}{e^{(\varepsilon-\mu)/k_BT} + 1}d\varepsilon$$

| approximate $E_c - \mu \gg k_BT$ and $\varepsilon > E_c$, so we write $\dfrac{1}{e^{(\varepsilon-\mu)/k_BT} + 1} \approx e^{-(\varepsilon-\mu)/k_BT}$ (4.6)

| also put in $1 = e^{E_c/k_BT}e^{-E_c/k_BT}$

$$n_c \approx \frac{1}{V}\int_{E_c}^{\infty} D(\varepsilon)e^{E_c/k_BT}e^{-E_c/k_BT}e^{-(\varepsilon-\mu)/k_BT}d\varepsilon \tag{4.7}$$

$$n_c = \frac{1}{V}e^{-(E_c-\mu)/k_BT}\int_{E_c}^{\infty} e^{-(\varepsilon-E_c)/k_BT}D(\varepsilon)d\varepsilon$$

| since the electrons occupy the bottom of the CB first,

| we approximate a parabola for the CB $\varepsilon_{\vec{k}} = E_c + \dfrac{\hbar^2 k^2}{2m_c^*}$ (4.8)

| then the DOS is $D(\varepsilon) = \dfrac{V}{2\pi^2}\left(\dfrac{2m_c^*}{\hbar^2}\right)^{3/2}\sqrt{\varepsilon - E_c}$

$$n_c \approx e^{-(E_c-\mu)/k_BT} \frac{1}{2\pi^2}\left(\frac{2m_c^*}{\hbar^2}\right)^{3/2} \int_{E_c}^{\infty} \sqrt{\varepsilon - E_C}\, e^{-(\varepsilon-E_c)/k_BT} d\varepsilon$$

(4.9)

| let $\dfrac{\varepsilon - E_c}{k_BT} = \varepsilon'$ and use integral identity $\displaystyle\int_0^{\infty} \sqrt{x}\,e^{-x}dx = \frac{\sqrt{\pi}}{2}$

$$n_c = e^{-(E_c-\mu)/k_BT} \frac{1}{2\pi^2}\left(\frac{2m_c^*}{\hbar^2}\right)^{3/2} \int_0^{\infty} \sqrt{k_BT}\,\sqrt{\varepsilon'}\, e^{-\varepsilon'} k_BT \, d\varepsilon'$$

(4.10)

$$n_c = e^{-(E_c-\mu)/k_BT} \frac{1}{2\pi^2}\left(\frac{2m_c^*}{\hbar^2}\right)^{3/2} (k_BT)3/2 \frac{\sqrt{\pi}}{2}$$

(4.11)

$$n_c = 2\left(\frac{m_c^* k_BT}{2\pi\hbar^2}\right)^{3/2} e^{-\frac{E_c-\mu}{k_BT}},$$

(4.12)

which is the expression for the concentration of electrons in the CB. The corresponding expression for holes in the VB is $p_v = 2\left(\dfrac{m_v^* k_BT}{2\pi\hbar^2}\right)^{3/2} e^{-\frac{\mu-E_v}{k_BT}}$, which you will check in the exercises.

We now proceed to calculate the chemical potential as a function of temperature.

$$n_c = 2\left(\frac{m_c^* k_BT}{2\pi\hbar^2}\right)^{3/2} e^{\frac{\mu-E_c}{k_BT}}$$

(4.13)

| we use shorthand and let $N_c(T) = 2\left(\dfrac{m_c^* k_BT}{2\pi\hbar^2}\right)^{3/2}$ and $\beta = \dfrac{1}{k_BT}$

$$n_c = N_c(T)e^{\beta(\mu-E_c)}$$

(4.14)

Similarly,

$$p_v = 2\left(\frac{m_v^* k_BT}{2\pi\hbar^2}\right)^{3/2} e^{\frac{E_v-\mu}{k_BT}}$$

(4.15)

| we use shorthand and let $N_v(T) = 2\left(\dfrac{m_v^* k_BT}{2\pi\hbar^2}\right)^{3/2}$ and $\beta = \dfrac{1}{k_BT}$

$$p_v = N_v(T)e^{\beta(E_v-\mu)}$$

(4.16)

The law of mass action means that the product $n_c p_v$ is independent of the chemical potential![2]

---

[2] The law of mass action is true for non-degenerate semiconductors. Doping (or adding impurities) affects the chemical potential μ but this product (or law) is unaffected by doping and can therefore be used for doped semiconductors. 'Degenerate' semiconductors refers to heavily doped semiconductors such that the chemical potential is within the conduction band. A degenerate semiconductor behaves more like a metal.

$$n_c p_v = 4 \left(\frac{k_B T}{2\pi \hbar^2} \right)^3 (m_c^* m_v^*)^{3/2} e^{-\beta(E_c - E_v)} = N_c(T) N_v(T) e^{-\beta E_g} \tag{4.17}$$

For intrinsic semiconductors, $n_c = p_v$. We change their notation to $n_c = n_i$ and $p_v = p_i$, then $n_i = p_i$ and we have

$$n_i^2 = N_c(T) N_v(T) e^{-\beta E_g} \tag{4.18}$$

$$n_i = p_i = \sqrt{N_c(T) N_v(T)} \, e^{-\beta E_g / 2} \tag{4.19}$$

$$= 2 \left(\frac{k_B T}{2\pi \hbar^2} \right)^{3/2} (m_c^* m_v^*)^{3/4} e^{-\beta E_g / 2}. \tag{4.20}$$

We can now actually solve for the (intrinsic) chemical potential as a function of temperature.

$$n_i = p_i \implies 2 \left(\frac{m_c^* k_B T}{2\pi \hbar^2} \right)^{3/2} e^{\beta(\mu_i - E_c)} = 2 \left(\frac{m_v^* k_B T}{2\pi \hbar^2} \right)^{3/2} e^{\beta(E_v - \mu_i)} \tag{4.21}$$

$$\left(\frac{m_c^*}{m_v^*} \right)^{3/2} = e^{\beta(-2\mu_i + E_c + E_v)} \tag{4.22}$$

$$| \text{ write } E_g = E_c - E_v$$

$$k_B T \frac{3}{2} \ln \left(\frac{m_c^*}{m_v^*} \right) = -2\mu_i + E_g + 2E_v \tag{4.23}$$

$$\mu_i = E_v + \frac{1}{2} E_g + \frac{3}{4} k_B T \ln \left(\frac{m_v^*}{m_v^*} \right) \tag{4.24}$$

So μ_i sits in the middle of the bandgap when $m_v^* = m_c^*$ (whatever the temperature) or when the temperature is zero. But note that $\mu_i - E_v \gg k_B T$ and $E_c - \mu_i \gg k_B T$. Let us look at some numbers; typically, $\mu_i - E_v \approx E_c - \mu_i \approx E_g \approx 1 \, eV$, then $T = 300$ K and $k_B T \approx \frac{1}{40}$ eV. Therefore, this means that μ_i is never really too far from the middle of the bandgap.[3]

4.1.2 Extrinsic semiconductors

'Extrinsic' semiconductors refers to semiconductors whose properties are modified by adding impurities to them. There are two types of 'useful' impurities: donor

---

[3] Be careful of the use of words! The symbol μ is the chemical potential but some books call it the Fermi level. The Fermi level refers to the highest filled level in metals. In semiconductors, there is no energy level in the gap and when $T = 0$, it is also not the highest filled level. So we shall stick with 'chemical potential' in the case of semiconductors.

impurities, which donate electrons, and acceptor impurities, which accept electrons (and 'annihilate' the electrons).

4.1.2.1 Donor and acceptor impurities

Consider the intrinsic semiconductor Si, which has four valence electrons (it is in group four in the periodic table) and the donor impurity[4] phosphorus (P), which has five valence electrons (it is in group five of the periodic table). Then four electrons of P are used in the covalent bond and we can think of the fifth electron as being bounded by P^+. The model for this bound state can be treated as a modified hydrogen atom. The Hamiltonian is

$$H = \frac{p^2}{2m_c^*} - \frac{e^2}{4\pi\varepsilon_0\varepsilon_r r} \tag{4.25}$$

- where m_c^* is the conduction band effective mass as an approximation for the fifth electron.
- where ε_r is the dielectric constant of the medium. In this case, the fifth electron is 'immersed' in the Si medium ($\varepsilon_r = 11.7$).

Thus, to get the ground-state energy eigenvalue (for the donor atom P), we merely modify the constants in the hydrogen ground-state energy eigenvalue.

$$\text{original hydrogen energy eigenvalues: } E_n = -\frac{m_e}{2\hbar^2}\left(\frac{e^2}{4\pi\varepsilon_0}\right)^2\frac{1}{n^2} \tag{4.26}$$

$$\text{set } n = 1 \text{ for ground state: } E_1 = -\frac{m_e}{2\hbar^2}\left(\frac{e^2}{4\pi\varepsilon_0}\right)^2 \tag{4.27}$$

$$\text{modify } m_e \to m_c^* \text{ and put in } \varepsilon_r \text{ beside } \varepsilon_0: E_{\text{donor}} = -\frac{1}{\varepsilon_r^2}\frac{m_c^*}{m_e}\frac{m_e}{2\hbar^2}\left(\frac{e^2}{4\pi\varepsilon_0}\right)^2 \tag{4.28}$$

- So $\frac{1}{\varepsilon_r^2} = \frac{1}{11.7^2}$ and $\frac{m_c^*}{m_e} \approx 0.2$. Note that $\frac{m_e}{2\hbar^2}\left(\frac{e^2}{4\pi\varepsilon_0}\right)^2 = 13.6\text{ eV}$ so $E_{\text{donor}} \approx 0.02$ eV. See figure 4.11, which shows the donor energy level to be just below the bottom of the CB. Thus, a small amount of thermal energy can excite the 5th electron from the donor level into the CB.
- The corresponding Bohr radius for the donor electron is $a_{\text{donor}} = \varepsilon_r\frac{m_e}{m_c^*}\frac{4\pi\varepsilon_0\hbar^2}{m_e e^2}$ and $\frac{4\pi\varepsilon_0\hbar^2}{m_e e^2} = 0.53\text{Å}$, so $a_{\text{donor}} \approx 31\text{Å}$. This means that the impurity orbit is large and spans many atoms.

---

[4] Here we only consider *shallow impurities*, which are easily ionised by thermal energy. Another type is called *deep impurities*; its energy levels are far from the top of the VB or bottom of the CB. Such impurities thus trap carriers and they are also called *deep traps*.

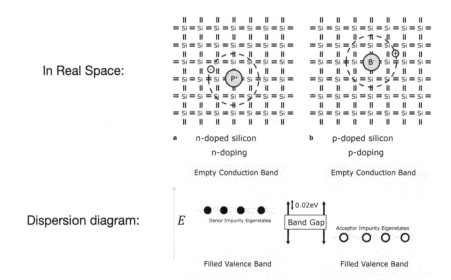

Figure 4.11. A doped semiconductor in real space and its effect on the energy dispersion. LEFT: doping with donor impurities. RIGHT: doping with acceptor impurities. Reprinted by permission from Springer Nature [1], Copyright (2009).

If we now consider boron (B) impurities in Si, three electrons from B are used in the covalent bond and one covalent bond is incomplete—this is a hole. This can be viewed as a hole bound to a B^-. This is called an acceptor state, and it is a hole state just above the valence band using similar arguments to those given above.

4.1.2.2 General properties of doped semiconductors: some relations

We now consider a semiconductor doped with donors (n-type). Let us derive an expression that relates the extrinsic carrier concentration n_c and $\Delta n = n_c - p_v$, noting that now $\Delta n \neq 0$. Even though the chemical potential $\mu \neq \mu_i$, the law of mass action still holds:[5] $n_c p_v = n_i^2$, thus

$$n_c - p_v = \Delta n$$
$$| \text{ then use } p_v = n_i^2/n_c$$

$$\quad (4.29)$$

$$n_c - \frac{n_i^2}{n_c} = \Delta n \quad (4.30)$$

$$n_c^2 - n_c \Delta n - n_i^2 = 0 \quad (4.31)$$

---

[5] Because we assume that the doped semiconductor is not degenerate.

$$n_c = \frac{\Delta n}{2} + \sqrt{\left(\frac{\Delta n}{2}\right)^2 + n_i^2}. \tag{4.32}$$

The positive square root solution is chosen because donors increase the n carrier concentration. We can now derive another relation for the extrinsic carrier concentration in terms of the chemical potential μ after doping. We can write the expression for the extrinsic carrier concentration in a similar form but in terms of the chemical potential μ:

$$n_c = 2\left(\frac{m_c^* k_B T}{2\pi \hbar^2}\right)^{3/2} e^{\beta(\mu - E_c)}$$

$$| \text{ recall that } n_i = 2\left(\frac{m_c^* k_B T}{2\pi \hbar^2}\right)^{3/2} e^{\beta(\mu_i - E_c)} \tag{4.33}$$

$$n_c = n_i e^{\beta(\mu - \mu_i)}. \tag{4.34}$$

$$\text{Similarly, } \quad p_v = p_i e^{-\beta(\mu - \mu_i)} \tag{4.35}$$

$$\text{take the difference, } \Delta n = n_c - p_v = n_i e^{\beta(\mu - \mu_i)} - p_i e^{-\beta(\mu - \mu_i)}$$

$$| \text{ recall that } p_i = n_i \tag{4.36}$$

$$\Delta n = 2n_i \sinh(\beta(\mu - \mu_i)) \tag{4.37}$$

$$\frac{\Delta n}{n_i} = 2 \sinh\left(\frac{\mu - \mu_i}{k_B T}\right). \tag{4.38}$$

4.1.2.3 The general properties of doped semiconductors: the dependence of carrier concentration on temperature

The behaviour of extrinsic semiconductors in relation to temperature can be divided into three regions (figure 4.12):

Freeze out: at low enough temperatures, the electrons in the CB fall to the donor level and the conductivity becomes essentially zero. The bottom of the CB falls to donor level ≈ 0.02 eV $\Longrightarrow T \approx 145$ K.

Extrinsic: the carriers are mostly provided by the dopants. This happens when the temperature $T \gtrsim 145$ K and thus essentially all the dopants are ionised.

Intrinsic: The carriers are provided by the dopants and also by thermally induced transitions from the VB to the CB (intrinsic carriers). Intrinsic carriers outnumber the carriers provided by dopants. Recall that intrinsic carriers $n_i \sim T^{3/2} e^{-\beta E_g}$, so the intrinsic carrier concentration increases rapidly with increasing temperature.

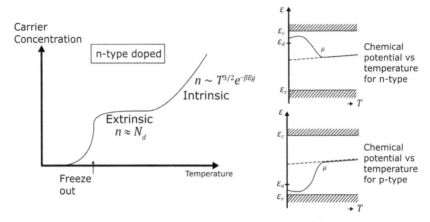

Figure 4.12. LEFT: the behaviour of extrinsic semiconductors in relation to temperature. RIGHT: the variation of chemical potential with temperature for both types of impurities. Reprinted by permission from Springer Nature [4], Copyright (2009).

4.1.2.4 The average occupancy of impurity levels

Consider the average occupancy of a single donor state using the grand canonical average. The energy of the donor state is denoted by ε_d. The average occupancy is

$$\langle n_d \rangle = \frac{\sum_j N_j e^{-\beta(E_j - \mu N_j)}}{\sum_j e^{-\beta(E_j - \mu N_j)}}. \tag{4.39}$$

Possible values of N_j (the number of electrons for the jth possibility) are:

$N_j = 0$: the donor atom is empty, the energy contribution is zero.

$N_j = 1$: the donor atom is occupied by an electron of spin ↑, the energy contribution is ε_d.

$N_j = 1$: the donor atom is occupied by an electron of spin ↓, the energy contribution is ε_d.

$N_j = 2$: the donor atom is occupied by two electrons (one of spin ↑ and one of spin ↓), the energy contribution is $2\varepsilon_d$. This case is too energy costly due to repulsion.

Hence, the average occupancy is calculated as

$$\langle n_d \rangle = \frac{0 + e^{-\beta(\varepsilon_d - \mu)} + e^{-\beta(\varepsilon_d - \mu)} + 2e^{-\beta(2\varepsilon_d - 2\mu)}}{e^0 + e^{-\beta(\varepsilon_d - \mu)} + e^{-\beta(\varepsilon_d - \mu)} + e^{-\beta(2\varepsilon_d - 2\mu)}} \tag{4.40}$$

| as mentioned above, we shall ignore the last term in both numerator and denominator

$$\approx \frac{2e^{-\beta(\varepsilon_d - \mu)}}{1 + 2e^{-\beta(\varepsilon_d - \mu)}} \tag{4.41}$$

$$= \frac{1}{\frac{1}{2}e^{\beta(\varepsilon_d - \mu)} + 1} \tag{4.42}$$

Similarly, if the energy of an acceptor state is denoted by ε_a, the average occupancy of that state is

$$\langle p_a \rangle = \frac{1}{\frac{1}{2}e^{\beta(\mu - \varepsilon_a)} + 1}. \tag{4.43}$$

4.1.2.5 Thermal equilibrium in a doped semiconductor

Consider a semiconductor that is doped with both types of impurities but $N_d > > N_a$, so it is an *n*-type semiconductor. At $T = 0$, we have the following occupation (figure 4.13):

- $n_c = 0$: no electrons in the CB
- $p_v = 0$: no holes in the VB
- $p_a = 0$: no holes bound to acceptor atoms or no holes at the acceptor level/band
- $n_d = N_d - N_a$: electrons bound to donor atoms. Subtraction is used because electrons are 'annihilated' by N_a holes. See figure 4.13.

At a finite temperature, $T \neq 0$, we have (figure 4.14):

- $n_c = N_c(T)e^{\beta(\mu - E_c)}$
- $p_v = N_v(T)e^{\beta(E_v - \mu)}$
- $n_d = N_d\langle n_d \rangle = \dfrac{N_d}{\frac{1}{2}e^{\beta(\varepsilon_d - \mu)} + 1}$
- $p_a = N_a\langle p_a \rangle = \dfrac{N_a}{\frac{1}{2}e^{\beta(\mu - \varepsilon_a)} + 1}$

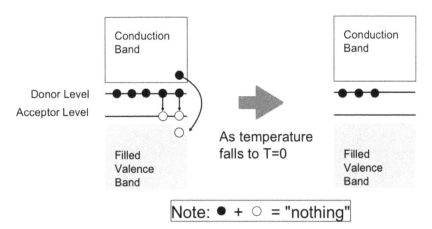

Figure 4.13. The occupation towards $T = 0$.

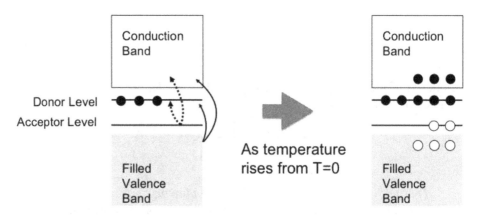

Figure 4.14. The occupation when the temperature rises from $T = 0$.

We also have the charge neutrality condition:[6]

$$n_c + n_d = N_d - N_a + p_v + p_a \qquad (4.44)$$

Thus, there are five equations and five unknowns: n_c, p_v, n_d, p_a, and μ. These five unknowns can be solved once we input the doping concentrations N_d and N_a, the energies E_c, E_v, ε_d, and ε_a, the temperature T, and the effective masses m_c^* and m_v^*.

4.2 Inhomogeneous semiconductors (towards devices)

We are now ready to proceed to a discussion of inhomogeneous semiconductors, which involve two or more extrinsic semiconductors joined together. This synthesis is important as it results in 'tunable' electron flow, which leads to the formation of electronic devices.

4.2.1 The *p*–*n* junction or diode

The simplest inhomogeneous semiconductor is simply a *p*-type extrinsic semiconductor joined to an *n*-type semiconductor. This results in a device called a 'diode'. We will investigate the properties of a *p*–*n* junction to see how the diode device operates.

4.2.1.1 The p–n junction: internal potential

Consider a piece of *n*-type semiconductor with a donor concentration N_d and another piece of *p*-type with an acceptor concentration N_a. Assume that they are lined up in the *z*-direction with the junction at $z = 0$.

$$N_d(z) = N_d\theta(z) \text{ and } N_a(z) = N_a(1 - \theta(z)) \qquad (4.45)$$

---

[6] My personal way of understanding this condition is to first think of $T = 0$, then $n_d = N_d - N_a$. At $T \neq 0$, electrons jump and add to $n_c + n_d$ and holes are created in $p_v + p_a$.

For $z \gg a$ (where a is the atomic spacing), the chemical potential is close to the donor levels. For $z \ll -a$, the chemical potential is close to the acceptor levels. Thus, we have a simple picture of the p–n junction (figure 4.15).

There must be a built-in potential $\phi(z)$ caused by joining extrinsic semiconductors of different carriers with different initial Fermi levels (figure 4.16). The energies of the bottom of the CB and the top of the VB can be written as

$$E_c(z) = E_c - e\phi(z) \text{ and } E_v(z) = E_v - e\phi(z)). \tag{4.46}$$

We can write the variation of the concentration as

$$n_c(z) = N_c(T)e^{-\beta(E_c(z)-\mu)} = N_c(T)e^{-\beta(E_c-e\phi(z)-\mu)} \tag{4.47}$$

$$p_v(z) = N_v(T)e^{-\beta(\mu-E_v(z))} = N_v(T)e^{-\beta(\mu-(E_v-e\phi(z)))}. \tag{4.48}$$

We assume the important case of $N_d \gg n_i$ and $N_a \gg n_i$. This is the high-concentration limit.

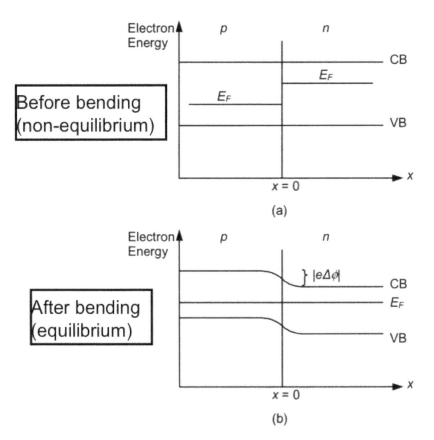

Figure 4.15. When two types of extrinsic semiconductors are 'joined', the differing chemical potentials result in a flow of carriers to achieve equilibrium. This creates an internal built-in potential difference $\Delta\phi$. Reprinted by permission from Springer Nature [5], Copyright (2007).

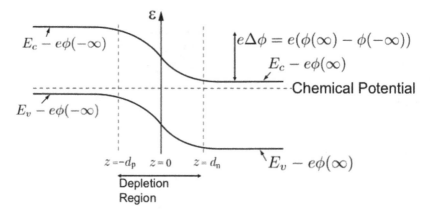

Figure 4.16. The built-in potential at equilibrium. Reprinted by permission from Springer Nature [6], Copyright (2018).

We assume that at the far ends of the semiconductors, the concentrations tend to N_d and N_a.

$$\lim_{z \to \infty} n_c(z) = N_c(T)e^{-\beta(E_c - e\phi(\infty) - \mu)} \approx N_d \tag{4.49}$$

$$\lim_{z \to -\infty} p_v(z) = N_v(T)e^{-\beta(\mu - E_v + e\phi(-\infty))} \approx N_a \tag{4.50}$$

We can solve for $e\Delta\phi = e(\phi(\infty) - \phi(-\infty))$. First, rewrite

$$-\beta(E_c - e\phi(\infty) - \mu) = \ln\left(\frac{N_d}{N_c(T)}\right) \tag{4.51}$$

$$-\beta(\mu - E_v + e\phi(-\infty)) = \ln\left(\frac{N_a}{N_v(T)}\right) \tag{4.52}$$

$$\Longrightarrow e\Delta\phi = e(\phi(\infty) - \phi(-\infty)) \tag{4.53}$$

$$e\Delta\phi = -\mu + E_c + k_B T \ln\left(\frac{N_d}{N_c(T)}\right) - \left[E_v - \mu + k_B T \ln\left(\frac{N_a}{N_v(T)}\right)\right] \tag{4.54}$$

$$e\Delta\phi = E_c - E_v + k_B T \ln\left(\frac{N_d}{N_c(T)}\right) + k_B T \ln\left(\frac{N_a}{N_v(T)}\right) \tag{4.55}$$

$$e\Delta\phi = E_g + k_B T \ln\left(\frac{N_d N_a}{N_c(T)N_v(T)}\right). \tag{4.56}$$

To solve for $\phi(z)$, we are supposed to use Poisson's equation $\vec{\nabla}^2\phi = -\frac{\rho}{\varepsilon_0}$. Here, we need its 1D version.

$$\frac{\partial^2 \phi(z)}{\partial z^2} = -\frac{\rho(z)}{\varepsilon_0 \varepsilon_r} \quad \text{with charge density: } \rho(z) = eN_d(z) + (-e)N_a(z) + (-e)n_c(z) + ep_v(z) \quad (4.57)$$

- The term $eN_d(z)$ appears because ionised donors are positively charged ions.
- The term $(-e)N_a(z)$ appears because ionised acceptors are negatively charged ions.
- The term $(-e)n_c(z)$ appears because carriers in the conduction band are negatively charged electrons.
- The term $ep_v(z)$ appears because carriers in the valence band are positively charged holes.

We may assume that all donors are ionised in $z > 0$ and all acceptors are ionised in $z < 0$, so

$$N_d(z) = N_d\theta(z) \text{ and } N_a(z) = N_a(1 - \theta(z)). \quad (4.58)$$

Recall that both $n_c(z) = N_c(T)e^{-\beta(E_c - \phi(z) - \mu)}$ and $N_d \approx N_c(T)e^{-\beta(E_c - e\phi(\infty) - \mu)}$,

$$\Longrightarrow n_c(z) = N_d e^{-\beta e(\phi(\infty) - \phi(z))} \text{ and similarly, } p_v(z) = N_a e^{-\beta e(\phi(z) - \phi(-\infty))}. \quad (4.59)$$

4.2.1.2 The p–n junction: the depletion region

Due to the flow of charges required to achieve equilibrium, an internal potential is set up (which means that an internal electric field is set up as well), and thus there is a region depleted of charges. We shall now calculate (with suitable approximations) the width of this region of neutrality.

Let us assume the following conditions:

$$\text{Region I: } \phi(z) = \phi(-\infty) \text{ for } z < -d_p \quad (4.60)$$

$$\text{Region II: } \phi(z) = \phi(\infty) \text{ for } z > d_n \quad (4.61)$$

$$\text{Region III: } \phi(z) \propto z \text{ for } -d_p < z < d_n \quad (4.62)$$

In Region I, substitute $\phi(z) = \phi(\infty)$ into equation (4.59) for p_v and get $p_v = N_a$. Equation (4.57) for charge density then gives $\rho_I(z) = ep_v - eN_a = 0$ because all the holes in the VB originate from the acceptor states.

In Region II, substitute $\phi(z) = \phi(\infty)$ into equation (4.59) for n_c and get $n_c = N_d$. Equation (4.57) for charge density then gives $\rho_{II}(z) = -en_c + eN_d = 0$ because the electrons in the CB are all due to ionised N_d.

In Region III, the potential $\phi(z)$ pushes the electrons and holes out, so $p_v(z) = 0 = n_c(z)$. Equation (4.57) becomes $\rho_{III}(z) = eN_d(z) - eN_a(z)$. Equation (4.58) then says

- $\rho_{III}(0 < z < d_n) = eN_d$
- $\rho_{III}(-d_p < z < 0) = -eN_a$

In the region $0 < z < d_{\mathrm{n}}$, Poisson's equation is

$$\frac{\partial^2 \phi(z)}{\partial z^2} = -\frac{eN_d}{\varepsilon_0 \varepsilon_r} \tag{4.63}$$

$$\text{one integration} \implies \frac{\partial \phi(z)}{\partial z} = -\frac{eN_d}{\varepsilon_0 \varepsilon_r} z + C_1 \tag{4.64}$$

$$\text{second integration} \implies \phi(z) = -\frac{eN_d}{2\varepsilon_0 \varepsilon_r} z^2 + C_1 z + C_2 \tag{4.65}$$

We set the integration constant using conditions at $z = d_{\mathrm{n}}$: $\phi(z = d_{\mathrm{n}}) = \phi(\infty)$ and $\frac{\partial \phi(z)}{\partial z}\Big|_{z=d_{\mathrm{n}}} = 0$, then

$$-\frac{eN_d}{\varepsilon_0 \varepsilon_r} + C_1 = 0 \implies C_1 = \frac{eN_d}{\varepsilon_0 \varepsilon_r} d_{\mathrm{n}} \tag{4.66}$$

$$\text{and then, } \phi(\infty) = -\frac{eN_d}{2\varepsilon_0 \varepsilon_r} d_{\mathrm{n}}^2 + \frac{eN_d}{\varepsilon_0 \varepsilon_r} d_{\mathrm{n}}^2 + C_2 \tag{4.67}$$

$$C_2 = -\frac{eN_d}{2\varepsilon_0 \varepsilon_r} d_{\mathrm{n}}^2 + \phi(\infty)$$

$$\text{Finally, the potential, } \phi(z) = -\frac{eN_d}{2\varepsilon_0 \varepsilon_r} z^2 + \frac{eN_d}{\varepsilon_0 \varepsilon_r} d_{\mathrm{n}} z - \frac{eN_d}{2\varepsilon_0 \varepsilon_r} d_{\mathrm{n}}^2 + \phi(\infty) \tag{4.68}$$

$$\phi(z) = \phi(\infty) - \frac{eN_d}{2\varepsilon_0 \varepsilon_r}(z^2 - 2d_{\mathrm{n}} z + d_{\mathrm{n}}^2) \tag{4.69}$$

$$\phi(z) = \phi(\infty) - \frac{eN_d}{2\varepsilon_0 \varepsilon_r}(z - d_{\mathrm{n}})^2 \tag{4.70}$$

$$\text{Similarly, for } -d_{\mathrm{p}} < z < 0, \quad \phi(z) = \phi(-\infty) + \frac{eN_d}{2\varepsilon_0 \varepsilon_r}(z + d_{\mathrm{p}})^2. \tag{4.71}$$

The charge conservation condition is that the total charge (concentration × volume) has the same magnitude on each side of the depletion region.

$$N_d d_{\mathrm{n}} = N_a d_{\mathrm{p}} \quad \text{(note that their cross sectional areas are the same)} \tag{4.72}$$

Note that the charge conservation condition is already included in the theory (as it should be) in the continuity condition of $\frac{\partial \phi(z)}{\partial z}$ at $z = 0$:

$$\frac{\partial \phi(\text{n-side})}{\partial z}\Big|_{z=0} = \frac{\partial \phi(\text{p-side})}{\partial z}\Big|_{z=0} \tag{4.73}$$

$$-\frac{eN_d}{\varepsilon_0\varepsilon_r}(-d_n) = \frac{eN_d}{\varepsilon_0\varepsilon_r}(d_p) \tag{4.74}$$

$$N_d d_n = N_a d_p \tag{4.75}$$

which is, of course, the same as equation (4.72).

We also need the continuity of $\phi(z)$ at $z = 0$ due to the usual assumption that potentials are continuous—even though they may be piecewise. This is a new condition that we impose now.

$$\phi(\text{n-side})|_{z=0} = \phi(\text{p-side})|_{z=0} \tag{4.76}$$

$$\phi(\infty) - \frac{eN_d}{2\varepsilon_0\varepsilon_r}(0 - d_n)^2 = \phi(-\infty) + \frac{eN_a}{2\varepsilon_0\varepsilon_r}(0 + d_p)^2 \tag{4.77}$$

$$\phi(\infty) - \phi(-\infty) = \frac{eN_d}{2\varepsilon_0\varepsilon_r}d_n^2 + \frac{eN_a}{2\varepsilon_0\varepsilon_r}d_p^2 \tag{4.78}$$

$$\Delta\phi = \frac{e}{2\varepsilon_0\varepsilon_r}\left(N_d d_n^2 + N_a d_p^2\right) \tag{4.79}$$

| then combine with charge conservation $N_d d_n = N_a d_p$

$$\Delta\phi = \frac{e}{2\varepsilon_0\varepsilon_r}\left(N_d d_n^2 + N_a\left(\frac{N_d}{N_a}d_n\right)^2\right) \tag{4.80}$$

$$\Longrightarrow d_n^2 = \frac{2\dfrac{N_a}{N_d}\varepsilon_0\varepsilon_r\Delta\phi}{e(N_a + N_d)} \tag{4.81}$$

$$\text{and, } d_p^2 = \frac{N_d^2}{N_a^2}d_n^2 = \frac{2\dfrac{N_d}{N_a}\varepsilon_0\varepsilon_r\Delta\phi}{e(N_a + N_d)} \tag{4.82}$$

Equal doping concentrations represent a special case: $N_a = N_d = N$, then

$$d_p = d_n = d = \sqrt{\frac{\varepsilon_0\varepsilon_r\Delta\phi}{eN}}$$

| recall equation (4.56), $e\Delta\phi = E_g + k_B T \ln\left(\dfrac{N^2}{N_c(T)N_v(T)}\right)$ $\quad$ (4.83)

| so we approximate $e\Delta\phi \approx E_g$

$$d = \sqrt{\frac{\varepsilon_0\varepsilon_r E_g}{e^2 N}}. \tag{4.84}$$

4.2.1.3 The p–n junction: current–voltage characteristics

We now apply a voltage $|V| = |V_+ - V_-|$, where V_+ is the higher-potential side of the battery and V_- is the lower side, across the p–n junction. We shall now determine the potential difference between both sides (figure 4.17).

- Forward bias is defined as $V_+ + \phi(-\infty)$ and $V_- + \phi(\infty)$, so the new potential difference is

$$\text{'n–side'} - \text{'p–side'} = V_- + \phi(\infty) - (V_+ + \phi(-\infty)) \tag{4.85}$$

$$\text{new potential difference} = \phi(\infty) - \phi(-\infty) - (V_+ - V_-) \tag{4.86}$$

$$\text{new potential difference} = \Delta\phi - V \quad \text{where } V > 0. \tag{4.87}$$

- Reverse bias is defined as $V_- + \phi(-\infty)$ and $V_+ + \phi(\infty)$, so

$$\text{new potential difference} = V_+ + \phi(\infty) - (V_- + \phi(-\infty)) \tag{4.88}$$

$$\text{new potential difference} = \phi(\infty) - \phi(-\infty) - (V_- - V_+) \tag{4.89}$$

$$\text{new potential difference} = \Delta\phi - V \quad \text{where } V < 0. \tag{4.90}$$

We now need to know how the depletion width changes with applied voltage. Recall equations (4.81) and (4.82), where $d_\mathrm{p}, d_\mathrm{n} \propto \sqrt{\Delta\phi} = \sqrt{\text{internal potential difference}}$. So we deduce

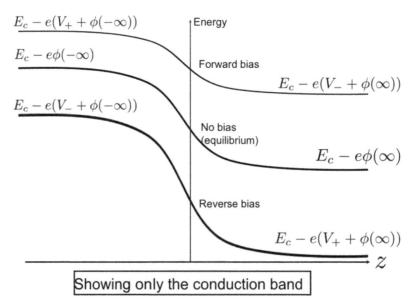

Figure 4.17. Changes to band bending due to the application of a bias.

$$d_p(V) \propto \sqrt{\text{new potential difference}} \tag{4.91}$$

$$d_p(V) \propto \sqrt{\Delta\phi - V} \tag{4.92}$$

$$d_p(V) \propto \sqrt{\Delta\phi}\sqrt{1 - \frac{V}{\Delta\phi}} \tag{4.93}$$

$$d_p(V) = d_p(0)\sqrt{1 - \frac{V}{\Delta\phi}}. \tag{4.94}$$

We proceed similarly for $d_n(V)$.

We can now calculate the current–voltage characteristics of the p–n junction under a bias. The electrical current density is contributed by both holes and electrons. So,

$$\vec{J}_{\text{total}} = e\vec{J}_h + (-e)\vec{J}_e \tag{4.95}$$

where $\vec{J}_h$ and $\vec{J}_e$ are the hole number and electron number current densities respectively.[7] From common sense, we know that when $V = 0$, there is no net current. Thus, there is actually a balancing carrier flow across the depletion region at $V = 0$. Consider the hole current, which has two components:

Generation current J_h^{gen}: current from the n side to the p side.

- On the n side, close to d_n, electrons are thermally excited and leave the VB, creating holes in the VB.
- These holes may drift into the depletion layer. If so, the strong E field in the depletion layer will sweep them to the p-side.
- This generation current is not sensitive to the applied voltage, as the internal electric field is quite strong.

Recombination current J_h^{rec}: current from the p side to the n side.

- Hole current flows from the p side to the n side due to the higher concentration on the p side.
- The electric field in the depletion region opposes this flow.
- Only holes with enough thermal energy can cross over and contribute to the current.
- Eventually, the holes recombine with electrons on then n side.
- So, we can assume a thermal relationship $|\vec{J}_h^{\text{rec}}| \propto e^{-e(\Delta\phi - V)/k_B T} = Ae^{-e(\Delta\phi - V)/k_B T}$.

$$\text{At } V = 0, \text{ we have } |\vec{J}_h^{\text{rec}}|_{V=0} = |\vec{J}_h^{\text{gen}}| \tag{4.96}$$

---

[7] The $(-e)$ then ensures that electron flow and conventional current match.

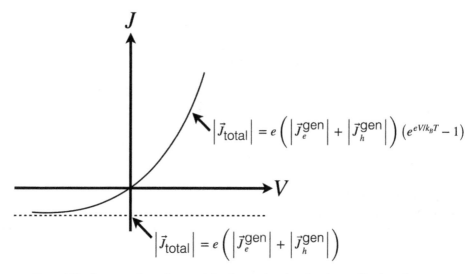

$$\left|\vec{J}_{\text{total}}\right| = e\left(\left|\vec{J}_e^{\,\text{gen}}\right| + \left|\vec{J}_h^{\,\text{gen}}\right|\right)(e^{eV/k_BT} - 1)$$

$$\left|\vec{J}_{\text{total}}\right| = e\left(\left|\vec{J}_e^{\,\text{gen}}\right| + \left|\vec{J}_h^{\,\text{gen}}\right|\right)$$

Figure 4.18. Current–voltage characteristic of a p–n junction showing rectification of current.

$$\Longrightarrow Ae^{-e\Delta\phi/k_BT} = |\vec{J}_h^{\,\text{gen}}| \tag{4.97}$$

$$\Longrightarrow |\vec{J}_h^{\,\text{rec}}| = |\vec{J}_h^{\,\text{gen}}|e^{eV/k_BT} \tag{4.98}$$

total hole current, $\vec{J}_h = \vec{J}_h^{\,\text{rec}} + \vec{J}_h^{\,\text{gen}}$

| note that these two currents are opposite
$$\tag{4.99}$$

$$= |\vec{J}_h^{\,\text{gen}}|e^{eV/k_BT}\hat{z} + |\vec{J}_h^{\,\text{gen}}|(-\hat{z}) \tag{4.100}$$

$$= |\vec{J}_h^{\,\text{gen}}|(e^{eV/k_BT} - 1)\hat{z}. \tag{4.101}$$

We can repeat the story for electrons and realise that the electron flow follows a similar expression but its directions are opposite (figure 4.18), so

$$\vec{J}_e = |\vec{J}_e^{\,\text{gen}}|(e^{eV/k_BT} - 1)(-\hat{z}) \tag{4.102}$$

$$\therefore \vec{J}_{\text{total}} = e\vec{J}_e + (-e)\vec{J}_h = e\left(\left|\vec{J}_e^{\,\text{gen}}\right| + \left|\vec{J}_h^{\,\text{gen}}\right|\right)(e^{eV/k_BT} - 1)\hat{z}. \tag{4.103}$$

4.2.1.4 A special case: the tunnel diode
- The tunnel diode is a p–n junction which is very heavily doped.
- Thus, its internal potential is so large that the bands bend and the conduction electrons on the n side are near the holes on the p side.
- Electrons from the n side tunnel over to annihilate the holes on the p side, resulting in an extra tunnelling current.

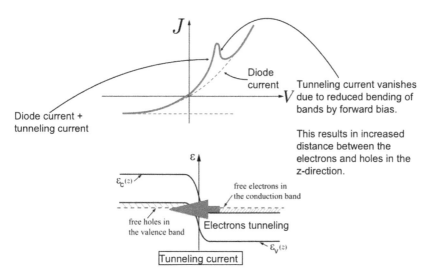

Figure 4.19. The tunnel diode. Reprinted by permission from Springer Nature [6], Copyright (2018).

- The tunnelling current increases with forward bias, as forward bias reduces the depletion region and thus increases the tunnelling transmission (figure 4.19).
- The tunnelling current is suddenly cut off when the forward bias reduces the band bending sufficiently that the conduction electrons (on the n side) are no longer near the holes (on the p side) (lower part of figure 4.19).

4.3 Exercises

4.3.1 Question: dispersion at the bottom of the CB

(Solution in section 6.4.1.) Write down the approximate parabolic dispersion at the bottom of the CB based on the constant energy surface (spheroid) of the semiconductor Si, as shown. To be specific, we take the top spheroid and assume that it is centred at some coordinate $k = (0, 0, 0.75\frac{2\pi}{a})$ (figure 4.20).

[Hints: You can start with a general expression for a triaxial ellipsoid. You need to introduce two types of effective mass because the spheroid has two symmetry axes.]

4.3.2 Question: the exciton energy level

(Solution in section 6.4.2.) The exciton is an electron–hole bound system. The electron has an effective mass of m_c^* and the hole has an effective mass of m_v^*. Look up quantum mechanics textbooks for central potential problems in which both masses must be accounted for (unlike the hydrogen atom problem, in which the nucleus is usually assumed to have infinite mass). Give the eigenenergy of the exciton by modifying the hydrogen atom eigenenergy result appropriately.

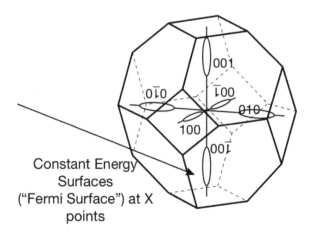

Figure 4.20. Constant energy surfaces of Si for question 4.3.1. Adapted with permission from [1], Copyright (2009) and [2], Copyright (2004), with permission from Springer Nature.

4.3.3 Question: intrinsic hole concentration in the VB

(Solution in section 6.4.3.) Derive, with suitable approximations, the intrinsic concentration of holes, p_v in the VB starting from $P_v = \frac{1}{V} \int_{-\infty}^{E_v} D(\varepsilon)(1 - f(\varepsilon))d\varepsilon$.

4.3.4 Question: some numbers to test

(Solution in section 6.4.4.) Let us use actual numbers to get a feeling for the carrier properties of semiconductors.

1. Assuming Si at 300 K and that the chemical potential is at the middle of the bandgap, evaluate the intrinsic electron concentration in the CB. You will need to search for the values of the Si bandgap and m_c^* on your own.
2. Now assume we have 1 mol of Si (at 300 K) and it has an impurity concentration of 0.01% of phosphorus. Assuming all the donor electrons are ionised, evaluate the ratio of donor-produced carriers to intrinsic carriers.
3. An intrinsic sample of GaAs is at a temperature of 300 K. Estimate the change in temperature required to increase its electrical conductivity by 10%. Use Drude electrical conductivity for this estimate.

Note: you may encounter an equation that you cannot solve algebraically. You may want to use a computing website such as http://www.wolframalpha.com to solve it numerically.

4.3.5 Question: impurity orbits in real space

(Solution in section 6.4.5.) Indium antimonide has $E_g = 0.23$ eV, relative permittivity $\varepsilon_r = 18$, and effective mass $m_c^* = 0.015m_e$.
1. Calculate the donor ionisation energy.
2. Calculate the radius of the ground-state orbit.

3. At what minimum donor concentration do adjacent impurity orbits start to overlap? You may assume that impurity atoms are distributed uniformly.[8]

4.3.6 Question: the internal electric field of a p–n junction

(Solution in section 6.4.6.) Sketch the internal electric potential and internal electric field in a p–n junction from $-\infty < z < \infty$ under the approximations used to calculate the depletion region in the lecture notes.

4.3.7 Question: the doping required for a tunnel diode

(Solution in section 6.4.7.) Calculate the doping concentration required for GaAs to become a tunnel diode that has an energy gap horizontal distance of 75 Å. Assume $N_d = N_a$ for simplicity.

References

[1] Ibach H and Lüth H 2009 *Solid-State Physics: An Introduction to Principles of Materials Science* (Berlin: Springer)

[2] Seeger K 2004 *Semiconductor Physics* 9th edn (Berlin: Springer)

[3] Peter Y and Cardona M 2010 *Fundamentals of Semiconductors: Physics and Materials Properties* 4th edn (Berlin: Springer)

[4] Sólyom J 2008 *Fundamentals of the Physics of Solids: Volume II: Electronic Properties* (Berlin: Springer)

[5] Patterson J D and Bailey B C 2007 *Solid-State Physics: Introduction to the Theory* 1st edn (Berlin: Springer)

[6] Quinn J J and Yi K-S 2018 *Solid State Physics: Principles and Modern Applications* 2nd edn (Berlin: Springer)

---

[8] This overlap tends to produce an impurity band: a band of energy levels which permit conductivity, presumably by a hopping mechanism in which electrons move from one impurity site to a neighbouring ionised impurity site.

IOP Publishing

Crystalline Solid State Physics
An interactive guide
Meng Lee Leek

Chapter 5

Magnetism

In the study of magnetic properties, the first thing that we must note is that magnetism is a quantum phenomenon. This is justified by the Bohr–van Leeuwen theorem.

We then look at how the magnetic material is affected by external magnetic fields. This is 'tracked' by the quantity called susceptibility. If the susceptibility is negative, the behaviour is called diamagnetism, otherwise the behaviour is called paramagnetism.

Finally, there is a wholly separate category of magnetism called collective magnetism. This is due to spin–spin interactions in solids, and its properties are based on the collective behaviour of many spins.

5.1 Preparation

5.1.1 Basic concepts and quantities

Here, we briefly review some concepts related to electromagnetism.

1. Magnetisation, $\vec{M}$:

$$\vec{M} = \lim_{\Delta V \to 0} \frac{1}{\Delta V} \sum_i^{\text{over } \Delta V} \vec{\mu}_i \tag{5.1}$$

where $\vec{\mu}_i$ is the magnetic moment of the ith atom and ΔV is a small volume.[1] We can see from this expression that magnetisation is also the magnetic dipole moment per unit volume.

2. $\vec{H}$-field:

The magnetisation current is $\vec{J}_M = \vec{\nabla} \times \vec{M}$ and the total $\vec{B}$-field is

---

[1] This definition is not very precise. In particular, what happens if the 'small volume' is about the size of an atom? Essentially, the microscopic (or local) aspect has to be treated carefully. We will not go into these details.

doi:10.1088/978-0-7503-5217-8ch5
© IOP Publishing Ltd 2023

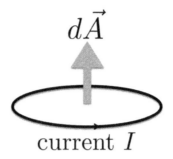

current I

Figure 5.1. A current loop and the area element vector.

$$\vec{\nabla} \times \vec{B} = \mu_0(\vec{J} + \vec{J}_M), \tag{5.2}$$

where $\vec{J}$ is the ordinary current. We define the $\vec{H}$-field as

$$\vec{H} = \frac{1}{\mu_0}\vec{B} - \vec{M} \implies \vec{\nabla} \times \vec{H} = \vec{J}. \tag{5.3}$$

3. Magnetic susceptibility, χ:
   Assume the constitutive equation $\vec{M} = \chi\vec{H}$; then

$$\vec{B} = \mu_0\left(\vec{H} + \vec{M}\right) = \mu_0\left(1 + \chi\right)\vec{H}. \tag{5.4}$$

4. Magnetic energy:
   The infinitesimal magnetic moment (figure 5.1): $d\vec{\mu} = Id\vec{A}$ (5.5)

$$\vec{\mu} = I \int d\vec{A} \tag{5.6}$$

$$E = -\vec{\mu} \cdot \vec{B} \quad \text{from electromagnetism} \tag{5.7}$$

The quantum Hamiltonian is inspired by that expression.
5. The Bohr magneton, μ_B:
   Consider hydrogen in the $n = 1$ ground state. The magnetic moment of the electron is (figure 5.2)

$$\mu = I \times A = \pi r^2 I \tag{5.8}$$

$$\mu = \pi r^2 \frac{-e}{t} \tag{5.9}$$

$$\mu = \pi r^2 \frac{-e}{2\pi r/v} \tag{5.10}$$

$$\mu = -\frac{evr}{2} \tag{5.11}$$

| use the Bohr–Sommerfeld quantization rule $m_e vr = \hbar$

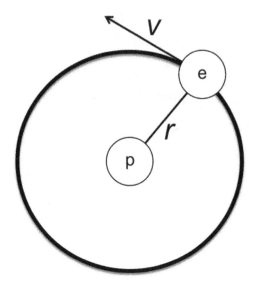

Figure 5.2. The hydrogen atom.

$$\mu = -\frac{e\hbar}{2m_e}$$

$$\tag{5.12}$$

| define the Bohr magneton $\mu_B = \dfrac{e\hbar}{2m_e}$

$$\mu = -\mu_B \tag{5.13}$$

5.1.2 Overview and classification

Magnetic phenomena can be classified into three main groups:

1. Diamagnetism
   - For diamagnetism, susceptibility $\chi^{\mathrm{dia}} < 0$.
   - An external field induces magnetic dipoles which orientate themselves antiparallel to the field following Lenz's law.
   - Thus, it is a property of all materials.
   - However, it is a weak effect that is usually overwhelmed by other magnetic effects.
   - Superconductors are perfect diamagnets, i.e. the external field is completely negated in the superconductor, or $\chi^{\mathrm{dia}} = -1$.
2. Paramagnetism
   - Susceptibility $\chi^{\mathrm{para}} > 0$.
   - Involves permanent magnetic dipoles that can occur in two ways:
     - Localised moments due to partially filled inner electron shells (Curie, Langevin paramagnetism).
     - Itinerant moments due to free conduction electrons (Pauli paramagnetism).

3. Collective or ordered magnetism
   - For collective or ordered magnetism, susceptibility is a complicated function $\chi = \chi(T, H, \text{history of material})$.
   - This type of magnetism is due to (quantum mechanical) exchange/interaction between the permanent magnetic dipoles.
   - There is a critical temperature T_C, below which spontaneous magnetisation occurs.
   - This type of magnetism has three main types: ferromagnetism, ferrimagnetism, and antiferromagnetism.

5.1.3 The Bohr–van Leeuwen theorem

Here, we will describe the Bohr–van Leeuwen theorem to justify the statement that magnetic phenomena are quantum in nature.

The Bohr–van Leeuwen theorem states:

'In a classical system, there is no thermal equilibrium magnetization.'

$\boxed{\text{Proof:}}$

Assume a 3D solid with N identical ions, each of which has a magnetic moment $\vec{\mu}$. The magnetisation is

$$\vec{M} = \frac{N}{V}\langle\vec{\mu}\rangle, \tag{5.14}$$

where $\langle\vec{\mu}\rangle$ is written using the classical statistical average,

$$\langle\vec{\mu}\rangle = \frac{\int dx_1\cdots dx_{3N}dp_1\cdots dp_{3N}\,\vec{\mu}\,e^{-\beta H}}{\int dx_1\cdots dx_{3N}dp_1\cdots dp_{3N}e^{-\beta H}}, \tag{5.15}$$

where $\beta = \frac{1}{k_B T}$ (k_B is the Boltzmann constant and T is the thermodynamic temperature). From $E = -\vec{\mu}\cdot\vec{B}$, we can write

$$\vec{\mu} = -\frac{\partial E}{\partial\vec{B}} \implies \vec{\mu} = -\frac{\partial H}{\partial\vec{B}}, \tag{5.16}$$

where H is the classical Hamiltonian. We can then write

$$\langle\vec{\mu}\rangle = \frac{\int dx_1\cdots dx_{3N}dp_1\cdots dp_{3N}\left(-\frac{\partial H}{\partial\vec{B}}\right)e^{-\beta H}}{\int dx_1\cdots dx_{3N}dp_1\cdots dp_{3N}e^{-\beta H}}$$

$$\tag{5.17}$$

| define the classical partition function $Z = \frac{1}{N!h^{3N}}\int dx_1\cdots dx_{3N}dp_1\cdots dp_{3N}e^{-\beta H}$

| note $\frac{\partial Z}{\partial\vec{B}} = \frac{1}{N!h^{3N}}\int dx_1\cdots dx_{3N}dp_1\cdots dp_{3N}\left(-\beta\frac{\partial H}{\partial\vec{B}}\right)e^{-\beta H}$, so we can write

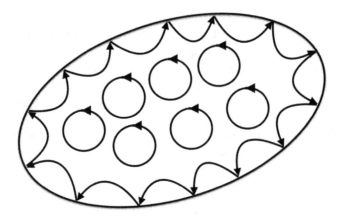

Figure 5.3. A sketch of Bohr–van Leeuwen theorem. The closed orbits and skipping orbits contribute opposite magnetic moments and so they cancel out.

$$\langle \vec{\mu} \rangle = \frac{1}{\beta Z} \frac{\partial Z}{\partial \vec{B}}. \tag{5.18}$$

The classical Hamiltonian with a magnetic field $\vec{B} = \vec{\nabla} \times \vec{A}$ takes the form

$$H = \frac{1}{2m} \sum_{i=1}^{3N} \left(\vec{p}_i + e\vec{A}_i \right)^2 + \text{other terms without } \vec{A} \tag{5.19}$$

$$Z = \frac{1}{N! h^{3N}} \int dx_1 \cdots dx_{3N} dp_1 \cdots dp_{3N} \exp \left(-\frac{\beta}{2m} \sum_{i=1}^{3N} (\vec{p}_i + e\vec{A}_i)^2 + \cdots \right) \tag{5.20}$$

| change variables $\vec{u}_i = \vec{p}_i + e\vec{A}_i$

| and note the integration limits for $\vec{p}_i$ goes from $-\infty$ to $+\infty$

$$Z = \frac{1}{N! h^{3N}} \int dx_1 \cdots dx_{3N} du_1 \cdots du_{3N} \exp \left(-\frac{\beta}{2m} \sum_{i=1}^{3N} \vec{u}_i^2 + \cdots \right), \tag{5.21}$$

which gives the result that Z is independent of $\vec{B}$. In fact, Z is the same as that without the magnetic field. Thus $\frac{\partial Z}{\partial \vec{B}} = 0$ and $\langle \vec{\mu} \rangle = 0$, hence we have proved that the average magnetisation at thermal equilibrium is always zero in the context of classical statistical mechanics.

We provide a sketch of the theorem. See figure 5.3.

5.2 Diamagnetism

5.2.1 General: an atom in a magnetic field

Assume there are Z (core) electrons in an atom in a magnetic field $\vec{B} = \vec{\nabla} \times \vec{A}$. The quantum Hamiltonian is as follows (where V is the atomic potential):

$$H = \sum_{i=1}^{Z} \frac{1}{2m_e} \left(\vec{p}_i + e\vec{A}_i \right)^2 + V \tag{5.22}$$

$$H = \sum_{i=1}^{Z} \frac{1}{2m_e} \left(\vec{p}_i^{\,2} + e(\vec{p}_i \cdot \vec{A}_i + \vec{A}_i \cdot \vec{p}_i) + e^2\vec{A}_i \cdot \vec{A}_i \right) + V$$

| choose Coulomb gauge $\vec{\nabla} \cdot \vec{A} = 0$ $\tag{5.23}$

| writing $\vec{A}(\vec{r}) = \frac{1}{2}(\vec{B} \times \vec{r})$ will satisfy both $\vec{\nabla} \cdot \vec{A} = 0$ and $\vec{B} = \vec{\nabla} \times \vec{A}$

$$H = \sum_{i=1}^{Z} \frac{1}{2m_e} \left(\vec{p}_i^{\,2} + e(\vec{p}_i \cdot \vec{A}_i + \vec{A}_i \cdot \vec{p}_i) + e^2\left(\frac{1}{2}(\vec{B} \times \vec{r}_i)\right)^2 \right) + V$$

| note that in position representation, using $\vec{\nabla}_i \cdot \vec{A}_i = 0$, $\tag{5.24}$

| write $\vec{p}_i \cdot \vec{A}_i + \vec{A}_i \cdot \vec{p}_i \rightarrow -i\hbar\vec{\nabla}_i \cdot (\vec{A}_i\psi) + \vec{A}_i \cdot (-i\hbar\vec{\nabla}_i\psi) = -2i\hbar\vec{A}_i \cdot \vec{\nabla}_i\psi \rightarrow 2\vec{A}_i \cdot \vec{p}_i$

| then substitute $\vec{A}_i = \frac{1}{2}(\vec{B} \times \vec{r}_i)$

$$H = \sum_{i=1}^{Z} \frac{1}{2m_e} \left(\vec{p}_i^{\,2} + e(\vec{B} \times \vec{r}_i) \cdot \vec{p}_i + \frac{e^2}{4}(\vec{B} \times \vec{r}_i)^2 \right) + V$$

| write $(\vec{B} \times \vec{r}_i) \cdot \vec{p}_i = (\vec{r}_i \times \vec{p}_i) \cdot \vec{B}$ which is the triple product identity $\tag{5.25}$

| and note that $\vec{r}_i \times \vec{p}_i = \hbar\vec{L}_i$ where $\vec{L}_i$ is the dimensionless orbital angular momentum

$$H = \sum_{i=1}^{Z} \frac{1}{2m_e} \left(\vec{p}_i^{\,2} + e\hbar\vec{L}_i \cdot \vec{B} + \frac{e^2}{4}(\vec{B} \times \vec{r}_i)^2 \right) + V$$

| include spin of the electrons interacting with the magnetic field $H_{\text{spin}} = \mu_B g_0 \vec{S} \cdot \vec{B}$ $\tag{5.26}$

| where $g_0 \approx 2$ and $\mu_B = \frac{e\hbar}{2m_e}$ is the Bohr magneton

$$H = \sum_{i=1}^{Z} \frac{\vec{p}_i^{\,2}}{2m_e} + V + \underbrace{\sum_{i=1}^{Z} \mu_B(\vec{L}_i + g_0\vec{S}_i) \cdot \vec{B}}_{\text{paramagnetism}} + \underbrace{\sum_{i=1}^{Z} \frac{e^2}{8m_e}(\vec{B} \times \vec{r}_i)^2}_{\text{diamagnetism}}. \tag{5.27}$$

We mentioned earlier that diamagnetism is a phenomena in all materials but the Hamiltonian seems to indicate that paramagnetism and diamagnetism are universal phenomena. Actually, the paramagnetism term vanishes for certain states of the atom. We can see this by first rewriting the paramagnetic Hamiltonian (we consider only one term in the sum and drop the index i for simplicity):

$H^{\text{para}} = \mu_B(\vec{L} + g_0\vec{S}) \cdot \vec{B}$

| assume $\vec{L} + g_0\vec{S} = k\vec{J}$ where k is some number. Then dot $\vec{J}$ on both sides. $\tag{5.28}$

| to get $\vec{J} \cdot (\vec{L} + g_0\vec{S}) = k\vec{J}^2 \Rightarrow k = \frac{1}{\vec{J}^2}\vec{J} \cdot (\vec{L} + g_0\vec{S}) \Rightarrow \vec{L} + g_0\vec{S} = \vec{J}\frac{\vec{J} \cdot (\vec{L} + g_0\vec{S})}{\vec{J}^2}$

$$H^{\text{para}} = \mu_B \vec{B} \cdot \vec{J} \frac{1}{\vec{J}^2}(\vec{J} \cdot \vec{L} + g_0 \vec{J} \cdot \vec{S})$$

| note $(\vec{J} - \vec{L})^2 = \vec{J}^2 + \vec{L}^2 - 2\vec{J} \cdot \vec{L}$ where $\vec{J} = \vec{L} + \vec{S}$ and $\vec{J} \cdot \vec{L}$ commutes with $\vec{L} \cdot \vec{J}$ (5.29)

| a similar expression is used for $\vec{J} \cdot \vec{S}$

$$H^{\text{para}} = \mu_B \vec{B} \cdot \vec{J} \frac{1}{2\vec{J}^2}\left(\vec{J}^2 + \vec{L}^2 - (\vec{J} - \vec{L})^2 + g_0(\vec{J}^2 + \vec{S}^2 - (\vec{J} - \vec{S})^2)\right)$$

(5.30)

| use $\vec{J} = \vec{L} + \vec{S}$

$$H^{\text{para}} = \frac{\mu_B \vec{B} \cdot \vec{J}}{2\vec{J}^2}\left(\vec{J}^2 + \vec{L}^2 - \vec{S}^2 + g_0\left(\vec{J}^2 + \vec{S}^2 - \vec{L}^2\right)\right)$$

| recall from QM that the squares of (dimensionless) angular momentum operators (5.31)

| have eigenvalues: $\vec{J}^2 = J(J + 1)$, $\vec{L}^2 = L(L + 1)$ and $\vec{S}^2 = S(S + 1)$

| this is written in atomic physics notation

$$H^{\text{para}} = \mu_B \vec{B} \cdot \vec{J}\left(\frac{1 + g_0}{2} + \frac{g_0 - 1}{2}\left(\frac{S(S + 1) - L(L + 1)}{J(J + 1)}\right)\right)$$

(5.32)

| recall that $g_0 \approx 2$

$$H^{\text{para}} = \mu_B \vec{B} \cdot \vec{J}\left(1 + \frac{J(J + 1) - L(L + 1) + S(S + 1)}{2J(J + 1)}\right)$$

(5.33)

$$H^{\text{para}} = g_{\text{eff}} \mu_B \vec{B} \cdot \vec{J}$$

(5.34)

where g_{eff} is called the Landé g-factor. And so in a state $|J = 0\rangle$ (which is determined by Hund's rules in atomic physics), there is no paramagnetism and thus diamagnetism is truly universal.

5.2.2 Diamagnetism of core electrons: Larmor or Langevin diamagnetism

We will consider the state $|J = 0\rangle$ since in other states, the presence of paramagnetism overwhelms diamagnetism. So H^{dia} is now the perturbation Hamiltonian. Recall that in quantum mechanics, under first-order Rayleigh–Schrödinger time-independent perturbation theory, the first-order energy correction is

$$E_n^{(1)} = \langle n^{(0)}|H_{\text{int}}|n^{(0)}\rangle$$

(5.35)

$$E_0^{(1)} = \left\langle 0 \left| \sum_{i=1}^{Z} \frac{e^2}{8m_e}(\vec{B} \times \vec{r_i})^2 \right| 0 \right\rangle$$

(5.36)

| where we wrote $|J = 0\rangle = |0\rangle$

| assume the magnetic field is in the $z - axis$ $\vec{B} = (0, 0, B)$

$$E_0^{(1)} = \sum_{i=1}^{Z} \frac{e^2}{8m_e} \left\langle 0 \left| (Bx_i - y_iB)\hat{z} \cdot (Bx_i - y_iB)\hat{z} \right| 0 \right\rangle \tag{5.37}$$

$$E_0^{(1)} = \sum_{i=1}^{Z} \frac{e^2 B^2}{8m_e} \langle 0|x_i^2 + y_i^2|0\rangle$$

| assume the state is spherically symmetric which is reasonable for $J = 0$ (5.38)

| then the expectation values $\langle x_i^2 \rangle = \langle y_i^2 \rangle = \frac{1}{3}\langle r_i^2 \rangle$

$$E_0^{(1)} = \frac{e^2 B^2}{12m_e} \sum_{i=1}^{Z} \langle 0|r_i^2|0\rangle \tag{5.39}$$

Recall the magnetic moment is $\vec{\mu} = -\frac{\partial E}{\partial \vec{B}}$, so the moment of this single atom is

$$\mu = -\frac{\partial E_0^{(1)}}{\partial B} = -\frac{e^2 B}{6m_e} \sum_{i=1}^{Z} \langle 0|r_i^2|0\rangle. \tag{5.40}$$

Assume a 3D solid of volume V with N identical such atoms; its magnetisation is (recall $\vec{M} = \frac{N}{V}\langle \vec{\mu} \rangle$)

$$M = \frac{N}{V}\mu = -\frac{N}{V}\frac{e^2 B}{6m_e} \sum_{i=1}^{Z} \langle 0|r_i^2|0\rangle. \tag{5.41}$$

Recall $\vec{M} = \chi \vec{H}$ and $\vec{H} = \frac{1}{\mu_0}\vec{B} - \vec{M}$. Since $M \ll H$, we can approximate $H \approx \frac{1}{\mu_0}B$, thus the susceptibility is

$$\chi = \frac{M}{H} \approx \frac{\mu_0 M}{B} \tag{5.42}$$

$$\chi = -\frac{N}{V}\frac{e^2 \mu_0}{6m_e} \sum_{i=1}^{Z} \langle 0|r_i^2|0\rangle. \tag{5.43}$$

A few comments are in order:

- A classical derivation gives a very similar expression but the Bohr–van Leeuwen theorem tells us that the classical derivation is likely to be misleading.
- The susceptibility seems to be temperature independent but this is because the derivation is not a statistical mechanical one. Experimentally, it turns out that the temperature dependence is weak.
- The negative sign in χ confirms that it is an effect that follows Lenz's law.
- Remember that any other magnetic effect is larger than atomic diamagnetism (except in a superconductor).

5.2.3 The diamagnetism of conduction electrons: Landau diamagnetism

In metals that contain a sea of mobile electrons, the application of a magnetic field results in a diamagnetic effect. The assumption is that there are Z mobile free electrons. The Hamiltonian is the same as before (without the atomic potential V_i)

$$H = \sum_{i=1}^{Z} \frac{1}{2m_e}\left(\vec{p}_i + e\vec{A}_i\right)^2, \tag{5.44}$$

and it turns out to be exactly solvable. We take the magnetic field to be in the z-direction, $\vec{B} = (0, 0, B)$. We choose $\vec{A} = (0, Bx, 0)$, which satisfies $\vec{\nabla} \times \vec{A} = \vec{B}$ and $\vec{\nabla} \cdot \vec{A}$ (Coulomb gauge).

$$H = \sum_{i=1}^{Z} \frac{1}{2m_e}(p_{ix}\hat{x} + p_{iy}\hat{y} + p_{iz}\hat{z} + eBx\hat{y}) \cdot (p_{ix}\hat{x} + p_{iy}\hat{y} + p_{iz}\hat{z} + eBx\hat{y}) \tag{5.45}$$

$$H = \sum_{i=1}^{Z} \frac{1}{2m_e}\left(p_{ix}^2 + p_{iz}^2 + \left(p_{iy} + eBx\right)^2\right) \tag{5.46}$$

| we can rewrite it to make it look more familiar; define $x_{i0} = \dfrac{-p_{iy}}{eB}$ and $\omega_c = \dfrac{eB}{m_e}$

$$H = \sum_{i=1}^{Z}\left(\frac{p_{ix}^2}{2m_e} + \frac{1}{2}m_e\omega_c(x_i - x_{i0})^2 + \frac{p_{iz}^2}{2m_e}\right) \tag{5.47}$$

The (time-independent) Schrödinger equation is $H\psi = E\psi$, and for position representation, we have the usual replacements: $p_{ix} \rightarrow -i\hbar\frac{\partial}{\partial x_i}$, $p_{iy} \rightarrow -i\hbar\frac{\partial}{\partial y_i}$, and $p_{iz} \rightarrow -i\hbar\frac{\partial}{\partial z_i}$. Considering only one electron (thus dropping index i),

$$\left[-\frac{\hbar^2}{2m_e}\frac{\partial^2}{\partial x^2} + \frac{1}{2}m_e\omega_c^2\left(x - \frac{i\hbar}{eB}\frac{\partial}{\partial y}\right)^2 - \frac{\hbar^2}{2m_e}\frac{\partial^2}{\partial z^2}\right]\psi = E\psi \tag{5.48}$$

write ψ in the separable form $\psi = e^{ik_z z}e^{ik_y y}\phi(x)|$ \tag{5.49}

$$\left[-\frac{\hbar^2}{2m_e}\frac{\partial^2}{\partial x^2} + \frac{1}{2}m_e\omega_c^2\left(x + \frac{\hbar k_y}{eB}\right)^2 + \frac{\hbar^2 k_z^2}{2m_e}\right]\phi(x) = E\phi(x) \tag{5.50}$$

$$\left[-\frac{\hbar^2}{2m_e}\frac{\partial^2}{\partial x^2} + \frac{1}{2}m_e\omega_c^2\left(x + \frac{\hbar k_y}{eB}\right)^2\right]\phi(x) = \left(E - \frac{\hbar^2 k_z^2}{2m_e}\right)\phi(x). \tag{5.51}$$

This is simply a harmonic oscillator problem (with a shifted potential); hence, the energy eigenvalues are

$$E - \frac{\hbar^2 k_z^2}{2m_e} = \left(n + \frac{1}{2}\right)\hbar\omega_c \tag{5.52}$$

$$E = \left(n + \frac{1}{2}\right) \hbar \omega_c + \frac{\hbar^2 k_z^2}{2m_e}. \tag{5.53}$$

Thus the electron is quantised in the x–y plane perpendicular to the field and it is a free particle in the z-direction, which is the direction of the field.

We will now outline the statistical mechanics calculation leading to the susceptibility. The (quantum) partition function for a canonical ensemble is calculated using

$$Z = \int dk_z \sum_n e^{-\frac{E}{k_B T}} = \int dk_z \sum_n e^{-\frac{1}{k_B T}\left(\left(n+\frac{1}{2}\right)\hbar\omega_c + \frac{\hbar^2 k_z^2}{2m_e}\right)}. \tag{5.54}$$

The Helmholtz free energy F is then given by

$$F = -k_B T \ln Z. \tag{5.55}$$

The magnetisation is obtained from

$$M = -\frac{\partial F}{\partial B} = -\frac{\partial F}{\partial \mu_0 H}. \tag{5.56}$$

Finally, the magnetic susceptibility is obtained from

$$\chi = \left(\frac{\partial M}{\partial H}\right)_{H=0}. \tag{5.57}$$

The result for Landau diamagnetism is

$$\chi_{\text{Landau}} = -\frac{N}{2V} \frac{\mu_0 \mu_B^2}{\varepsilon_F}, \tag{5.58}$$

where ε_F is the Fermi energy.

5.3 Paramagnetism

5.3.1 Atomic paramagnetism: Curie or Langevin paramagnetism

Consider an atom whose electrons are in states that result in a net angular momentum of J (integer or half integer).[2] Recall from equation (5.34) that the paramagnetic Hamiltonian is

$$H^{\text{para}} = g_{\text{eff}} \mu_B \vec{B} \cdot \vec{J}. \tag{5.59}$$

Take $\vec{B} = B\hat{z}$, so $\vec{B} \cdot \vec{J} = BJ_z$. Denoting atomic states by $|J, m_J\rangle$, the energy eigenvalues are

$$H^{\text{para}}|J, m_J\rangle = g_{\text{eff}} \mu_B B J_z |J, m_J\rangle = g_{\text{eff}} \mu_B B m_J |J, m_J\rangle. \tag{5.60}$$

Note that there is no $\hbar$, as J_z is defined to be dimensionless.

---

[2] This is determined using Hund's Rules in atomic physics.

The calculation leading to the susceptibility uses statistical mechanics. The (quantum) partition function for a canonical ensemble is

$$Z = \sum_{m_J=-J}^{J} e^{-\frac{E}{k_B T}} \tag{5.61}$$

$$Z = \sum_{m_J=-J}^{J} e^{-\frac{g_{\text{eff}}\mu_B B m_J}{k_B T}} \tag{5.62}$$

$$\mid \text{ and let } x = \frac{g_{\text{eff}}\mu_B B}{k_B T}$$

$$Z = \sum_{m_J=-J}^{J} e^{-x m_J} \tag{5.63}$$

$$Z = e^{-xJ} + e^{-x(J-1)} + \ldots + e^{x(J-1)} + e^{xJ} \tag{5.64}$$

$$Z = e^{-xJ}(1 + e^x + e^{2x} + \ldots + e^{x(2J-1)} + e^{x2J})$$

$\mid$ which is a finite geometric series with the first term $= 1$, common factor $= e^x$

$\mid$ so use $S_N = \sum_{k=0}^{N} r^k = \dfrac{1 - r^{N+1}}{1 - r}$ and there are $2J + 1$ terms, so $N + 1 = 2J + 1$ \tag{5.65}

$$Z = e^{-xJ}\frac{1 - e^{x(2J+1)}}{1 - e^x} \tag{5.66}$$

$$Z = \frac{e^{-xJ} - e^{x(J+1)}}{1 - e^x} \tag{5.67}$$

$$Z = \frac{e^{-xJ}e^{-x/2} - e^{x(J+1)}e^{-x/2}}{e^{-x/2} - e^{x/2}} \tag{5.68}$$

$$Z = \frac{e^{-\frac{x}{2}(2J+1)} - e^{\frac{x}{2}(2J+1)}}{e^{-\frac{x}{2}} - e^{\frac{x}{2}}} \tag{5.69}$$

$$\mid \text{ use } \sinh z = \frac{1}{2}(e^z - e^{-z})$$

$$Z = \frac{\sinh\left(\frac{x}{2}(2J + 1)\right)}{\sinh\left(\frac{x}{2}\right)}. \tag{5.70}$$

Magnetisation is calculated from

$$M = -\frac{\partial F}{\partial B} M = -\frac{\partial}{\partial B}(-k_B T \ln Z) \tag{5.71}$$

$$M = k_B T \frac{1}{Z} \frac{\partial Z}{\partial B}$$

$$\mid \text{ but recall } x = \frac{g_{\text{eff}} \mu_B B}{k_B T} \implies \frac{\partial x}{\partial B} = \frac{g_{\text{eff}} \mu_B}{k_B T} \tag{5.72}$$

$$M = g_{\text{eff}} \mu_B \frac{1}{Z} \frac{\partial Z}{\partial x} \tag{5.73}$$

$$M = g_{\text{eff}} \mu_B \frac{1}{Z} \frac{\sinh\left(\dfrac{x}{2}\right)}{\sinh\left(\dfrac{x}{2}(2J+1)\right)}$$

$$\times \left(\frac{\dfrac{2J+1}{2}\cosh\left(\dfrac{x}{2}(2J+1)\right)}{\sinh\left(\dfrac{x}{2}\right)} - \frac{\sinh\left(\dfrac{x}{2}(2J+1)\right)\cosh\left(\dfrac{x}{2}\right)}{2\sinh^2\left(\dfrac{x}{2}\right)} \right) \tag{5.74}$$

$$M = g_{\text{eff}} \mu_B J \left(\frac{2J+1}{2J}\coth\left(\frac{2J+1}{2J}xJ\right) - \frac{1}{2J}\coth\left(\frac{xJ}{2J}\right) \right)$$

$$\mid \text{ define the Brillouin function } B_J(y) = \frac{2J+1}{2J}\coth\left(\frac{2J+1}{2J}y\right) - \frac{1}{2J}\coth\left(\frac{y}{2J}\right) \tag{5.75}$$

$$M = g_{\text{eff}} \mu_B J B_J(xJ). \tag{5.76}$$

The susceptibility can be calculated using $\chi = \left(\dfrac{\partial M}{\partial H}\right)_{H=0} = \mu_0 \left(\dfrac{\partial M}{\partial B}\right)_{B=0}$ but this makes the algebra very messy. We will calculate two specific cases only.

- Case 1: The smallest angular momentum value of $J = \frac{1}{2}$.

$$B_J(xJ)\big|_{J=1/2} = \frac{2\left(\dfrac{1}{2}\right)+1}{2\left(\dfrac{1}{2}\right)}\coth\left(\frac{2\left(\dfrac{1}{2}\right)+1}{2\left(\dfrac{1}{2}\right)}\right) - \frac{1}{2\left(\dfrac{1}{2}\right)}\coth\left(\frac{x\left(\dfrac{1}{2}\right)}{2\left(\dfrac{1}{2}\right)}\right) \tag{5.77}$$

$$B_J(xJ)\big|_{J=1/2} = 2\coth x - \coth\left(\frac{x}{2}\right)$$

$$\mid \text{ use the addition identity; } \coth x = \frac{\coth^2\left(\dfrac{x}{2}\right)+1}{2\coth\left(\dfrac{x}{2}\right)} \tag{5.78}$$

$$B_J(xJ)|_{J=1/2} = \frac{1}{\coth\left(\dfrac{x}{2}\right)} \tag{5.79}$$

$$B_J(xJ)|_{J=1/2} = \tanh\left(\frac{x}{2}\right) \tag{5.80}$$

$$\therefore \ M = g_{\text{eff}}\mu_B J \tanh\left(\frac{x}{2}\right) \tag{5.81}$$

$$\chi = \mu_0\left(\frac{\partial M}{\partial B}\right)_{B=0} \chi = \mu_0\left(\frac{\partial M}{\partial x}\frac{\partial x}{\partial B}\right)_{B=0 \text{ or } x=0} \tag{5.82}$$

$$\chi = \mu_0\left(\frac{g_{\text{eff}}\mu_B \dfrac{1}{2}}{2}\text{sech}^2\left(\frac{x}{2}\right)\frac{g_{\text{eff}}\mu_B}{k_B T}\right)_{x=0} \tag{5.83}$$

$$\chi = \frac{g_{\text{eff}}^2\mu_0\mu_B^2}{4k_B T} \tag{5.84}$$

- Case 2: $x \ll 1$, which is the commonly encountered experimental condition. We Taylor expand $B_J(xJ)$ to the first order,

$$B_J(xJ) = \frac{2J+1}{2J}\coth\left(\frac{2J+1}{2J}xJ\right) - \frac{1}{2J}\coth\left(\frac{xJ}{2J}\right) \tag{5.85}$$

| recall the Taylor series of $\coth y = \dfrac{1}{y} + \dfrac{y}{3} + \cdots$ for $y \ll 1$

$$B_J(xJ) \approx \frac{2J+1}{2J}\left(\frac{2}{(2J+1)x} + \frac{(2J+1)x}{2\times 3}\right) - \frac{1}{2J}\left(\frac{2}{x} + \frac{x}{2\times 3}\right) \tag{5.86}$$

$$B_J(xJ) = \frac{1}{Jx} + \frac{(2J+1)^2 x}{12J} - \frac{1}{Jx} - \frac{x}{12J} \tag{5.87}$$

$$B_J(xJ) = \frac{4J^2 + 4J}{12J}x \tag{5.88}$$

$$B_J(xJ) = \frac{J+1}{3}x \tag{5.89}$$

$$\therefore \ M = g_{\text{eff}}\mu_B J \frac{J+1}{3}x \tag{5.90}$$

$$\chi = \mu_0 \left(\frac{\partial M}{\partial x} \frac{\partial x}{\partial B} \right)_{B=0 \text{ or } x=0} \tag{5.91}$$

$$\chi = \mu_0 \left(g_{\text{eff}}\mu_B \frac{J(J+1)}{3} \frac{g_{\text{eff}}\mu_B}{k_B T} \right)_{x=0} \tag{5.92}$$

$$\chi = \frac{\mu_0 g_{\text{eff}}^2 \mu_B^2 J(J+1)}{3k_B T} \tag{5.93}$$

This relation of $\chi \propto \frac{1}{T}$ is known as Curie's law of paramagnetism.

5.3.2 The paramagnetism of conduction electrons: Pauli paramagnetism

Conduction electrons contribute a paramagnetic moment because each electron has spin $\frac{1}{2}$. For simplicity, the derivation will be done by assuming $T = 0$.

We split the density of states of the conduction electrons into 'spin up' and 'spin down' components.

$$\rho(E) = \rho_\uparrow(E) + \rho_\downarrow(E) \tag{5.94}$$

The magnetic field is assumed to be in the positive z-direction $\vec{B} = B\hat{z}$ and $\uparrow \Rightarrow m_s = +\frac{1}{2}$ (parallel to $\vec{B}$) and $\downarrow \Rightarrow m_s = -\frac{1}{2}$ (antiparallel to $\vec{B}$).

When there is no magnetic field, there are equal numbers of $\uparrow$-electrons and $\downarrow$-electrons so $\rho_\uparrow(E) = \rho_\downarrow(E)$, which means there is zero net magnetisation (figure 5.4).

On application of the magnetic field $\vec{B} = B\hat{z}$ and considering only spin, the Hamiltonian of the problem is the paramagnetic Hamiltonian $H^{\text{para}} = g_0\mu_B \vec{S} \cdot \vec{B}$.

- For spin up electrons:

$$H^{\text{para}}| \uparrow \rangle = 2\mu_B \vec{S} \cdot \vec{B}|m_s = +\frac{1}{2}\rangle = 2\mu_B BS_z|m_s = +\frac{1}{2}\rangle = \mu_B B|m_s = +\frac{1}{2}\rangle \tag{5.95}$$

- For spin down electrons:

$$H^{\text{para}}| \downarrow \rangle = 2\mu_B \vec{S} \cdot \vec{B}|m_s = -\frac{1}{2}\rangle = 2\mu_B BS_z|m_s = -\frac{1}{2}\rangle = -\mu_B B|m_s = -\frac{1}{2}\rangle \tag{5.96}$$

Thus, spin up electrons gain energy $\mu_B B$ and spin down electrons lose energy $\mu_B B$. The densities of states are modified accordingly:

$$\rho_\uparrow(E) \longrightarrow \rho_\uparrow(E - \mu_B B) \text{ and } \rho_\downarrow(E) \longrightarrow \rho_\downarrow(E + \mu_B B): \tag{5.97}$$

Figure 5.5 shows the situation clearly.

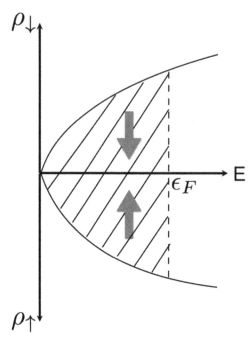

Figure 5.4. The density of states for 'free electrons' separated into spin up and spin down constituents. There are equal numbers of spin up and spin down electrons on average. The electrons are filled to ϵ_F, the Fermi energy.

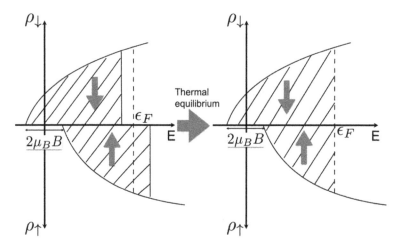

Figure 5.5. The densities of states for 'free electrons' separated into spin up and spin down parts under the influence of a magnetic field in the spin up direction. Spin up and spin down electrons gain $\mu_B B$ and lose $\mu_B B$, respectively. The system then reaches a new thermal equilibrium at ϵ_F.

There is now an excess of spin down electrons, which causes net magnetisation.

$$M = \frac{\mu_B}{V}(N_\downarrow - N_\uparrow) = \mu_B(n_\downarrow - n_\uparrow) \tag{5.98}$$

The number densities of the up and down electrons are denoted by $n_\uparrow$ and $n_\downarrow$ respectively. We need to evaluate the number densities.

$$n_\uparrow = \int_{-\infty}^{+\infty} dE \rho_\uparrow(E - \mu_B B) f^{FD}(E) \tag{5.99}$$

| where f^{FD} is the Fermi–Dirac distribution

$$n_\uparrow = \int_{\mu_B B}^{+\infty} dE \rho_\uparrow(E - \mu_B B) f^{FD}(E) \tag{5.100}$$

| make a change of variables $E' = E - \mu_B B$

$$n_\uparrow = \int_0^\infty dE' \rho_\uparrow(E') f^{FD}(E' + \mu_B B) \tag{5.101}$$

| since $\mu_B B \ll \varepsilon_F$, we Taylor expand $f^{FD}(E' + \mu_B B) \approx f^{FD}(E') + \mu_B B \frac{\partial f^{FD}(E')}{\partial E'}$

$$n_\uparrow \approx \int_0^\infty dE' \rho_\uparrow(E') \left(f^{FD}(E') + \mu_B B \frac{\partial f^{FD}(E')}{\partial E'} \right) \tag{5.102}$$

Similarly,

$$n_\downarrow \approx \int_0^\infty dE' \rho_\downarrow(E') \left(f^{FD}(E') - \mu_B B \frac{\partial f^{FD}(E')}{\partial E'} \right). \tag{5.103}$$

The magnetisation is

$$M = \mu_B(n_\downarrow - n_\uparrow)$$

$$M = \mu_B \left[\int_0^\infty dE'(\rho_\downarrow(E') - \rho_\uparrow(E')) f^{FD}(E') - \mu_B B \int_0^\infty dE'(\rho_\downarrow(E') + \rho_\uparrow(E')) \frac{\partial f^{FD}(E')}{\partial E'} \right] \tag{5.104}$$

| recall that $\rho_\downarrow(E) = \rho_\uparrow(E)$ and $\rho_\downarrow(E) + \rho_\uparrow(E) = \rho(E)$

$$M = -\mu_B^2 B \int_0^\infty dE' \rho(E') \frac{\partial f^{FD}(E')}{\partial E'}. \tag{5.105}$$

We can calculate the susceptibility now.

$$\chi_{\text{Pauli}} = \mu_0 \left(\frac{\partial M}{\partial B} \right)_{B=0} \tag{5.106}$$

$$\chi_{\text{Pauli}} = -\mu_0\mu_B^2 \int_0^\infty dE'\rho(E')\frac{\partial f^{FD}(E')}{\partial E'}$$

| at low temperature, we approximate $f^{FD}(E) \approx \theta(\varepsilon_F - E)$, the step function   (5.107)

| thus $\dfrac{\partial f^{FD}(E')}{\partial E'} \approx \dfrac{\partial}{\partial E'}\theta(\varepsilon_F - E') = -\delta(\varepsilon_F - E')$ and evaluate the integral

$$\chi_{\text{Pauli}} = \mu_0\mu_B^2\rho(\varepsilon_F)$$

| recall in free electron theory $\rho(\varepsilon_F) = \dfrac{3}{2}\dfrac{N}{V}\dfrac{1}{\varepsilon_F}$ (5.108)

$$\chi_{\text{Pauli}} = \frac{3N}{2V}\frac{\mu_0\mu_B^2}{\varepsilon_F} \tag{5.109}$$

Notice that there is a relationship between χ_{Landau} and χ_{Pauli}:

$$\chi_{\text{Landau}} = -\frac{1}{3}\chi_{\text{Pauli}}. \tag{5.110}$$

Thus, the total susceptibility of a metal in a magnetic field is

$$\chi_{\text{metal}} = \chi_{\text{Landau}} + \chi_{\text{Pauli}} + \chi_{\text{Larmor}} + \chi_{\text{Curie}} \tag{5.111}$$

$$\chi_{\text{metal}} = -\frac{1}{3}\chi_{\text{Pauli}} + \chi_{\text{Pauli}} + \chi_{\text{Larmor}} + \chi_{\text{Curie}} \tag{5.112}$$

$$\chi_{\text{metal}} = \frac{2}{3}\chi_{\text{Pauli}} + \chi_{\text{Larmor}} + \chi_{\text{Curie}}. \tag{5.113}$$

It turns out that the dominant contribution to χ_{metal} depends strongly on the material.

5.3.3 Van Vleck paramagnetism

Recall that in the discussion of diamagnetism, in the state $|J = 0\rangle = |0\rangle$, the paramagnetic Hamiltonian did not contribute to first-order perturbation, i.e.

$$\text{For } H_{\text{int}} = \sum_{i=1}^{Z}\mu_B(\vec{L}_i + g_0\vec{S}_i) \cdot \vec{B} + \frac{e^2}{8m_e}(\vec{B} \times \vec{r}_i)^2 \tag{5.114}$$

$$H_{\text{int}} = \sum_{i=1}^{Z}g_{\text{eff}}\mu_B\vec{B} \cdot \vec{J}_i + \frac{e^2}{8m_e}(\vec{B} \times \vec{r}_i)^2 \tag{5.115}$$

$$E_0^{(1)} = \langle 0|H_{\text{int}}|0\rangle \tag{5.116}$$

$$E_0^{(1)} = \sum_{i=1}^{Z} g_{\text{eff}} \mu_B \langle 0 | \vec{B} \cdot \vec{J}_i | 0 \rangle + \frac{e^2}{8m_e} \langle 0 | (\vec{B} \times \vec{r}_i)^2 | 0 \rangle \tag{5.117}$$

| and the first term is zero

$$E_0^{(1)} = \frac{e^2 B^2}{12 m_e} \sum_{i=1}^{Z} \langle 0 | r_i^2 | 0 \rangle. \tag{5.118}$$

From the second-order perturbation of $H^{\text{para}} = \sum_{i=1}^{Z} g_{\text{eff}} \mu_B \vec{B} \cdot \vec{J}_i$, we get the so-called 'Van Vleck paramagnetism', which we will discuss now.

Recall the second-order perturbation theory energy correction expression:

$$E_n^{(2)} = \sum_{m(\neq n)} \frac{|\langle n | H_{\text{int}} | m \rangle|^2}{E_n - E_m} \tag{5.119}$$

| in this case, the state for correction is $|n\rangle = |0\rangle$ and $H_{\text{int}} = H^{\text{para}}$

$$E_0^{(2)} = \sum_{m(\neq 0)} \frac{\left| \left\langle 0 \left| \sum_{i=1}^{Z} g_{\text{eff}} \mu_B \vec{B} \cdot \vec{J}_i \right| m \right\rangle \right|^2}{E_0 - E_m} \tag{5.120}$$

Note that the state $|J = 0\rangle = |0\rangle$ is usually the lowest energy state (cross-reference atomic physics); thus, $E_m - E_0 > 0$ and $E_0^{(2)} < 0$.

Again, take $\vec{B} = (0, 0, B)$, so $\vec{B} \cdot \vec{J}_i = B J_{iz}$.

$$E_0^{(2)} = \sum_{m(\neq 0)} \sum_{i=1}^{Z} g_{\text{eff}}^2 \mu_B^2 B^2 \frac{|\langle 0 | J_{iz} | m \rangle|^2}{E_0 - E_m} \tag{5.121}$$

The moment of this single atom is

$$\mu = -\frac{\partial E_0^{(2)}}{\partial B} = \sum_{m(\neq 0)} \sum_{i=1}^{Z} g_{\text{eff}}^2 \mu_B^2 B \frac{|\langle 0 | J_{iz} | m \rangle|^2}{E_m - E_0}. \tag{5.122}$$

The magnetisation of N identical atoms in volume V is $M = \frac{N}{V} \mu$. The susceptibility is

$$\chi \approx \frac{\mu_0 M}{B} = \frac{N}{V} \frac{\mu_0}{B} \mu \tag{5.123}$$

$$\chi = \frac{N}{V} \mu_0 \sum_{m(\neq 0)} \sum_{i=1}^{Z} g_{\text{eff}}^2 \mu_B^2 \frac{|\langle 0 | J_{iz} | m \rangle|^2}{E_m - E_0}. \tag{5.124}$$

This positive susceptibility is called Van Vleck paramagnetism and it is very small. It is temperature independent because this is a quantum mechanical derivation and not a statistical mechanical derivation. From a more detailed derivation, it turns out that the above expression for χ is essentially correct and that χ is temperature independent.

5.3.4 Isentropic demagnetisation or adiabatic demagnetisation

This technique is used to cool paramagnetic materials. Recall equation (5.90), the expression for magnetisation in atomic paramagnetism:

$$M = g_{\text{eff}}\mu_B \frac{J(J+1)}{3} x = g_{\text{eff}}^2 \mu_B^2 \frac{J(J+1)}{3} \frac{B}{k_B T}. \tag{5.125}$$

Thus, increasing the B-field increases the magnetisation and increasing T decreases M, which is compatible with common sense.

We will now describe some basic concepts in entropy in order to understand how it can be manipulated to achieve cooling.

- The Boltzmann definition of entropy is $S = k_B \ln \Omega$, where Ω is the number of (micro) arrangements.
- At low temperatures, all the magnetic moments align with the magnetic field and there is only one arrangement, $\Omega = 1$, therefore $S = 0$.
- At high temperatures or in a zero magnetic field, the system is disordered, and in state J, there are $\dim(m_J) = 2J + 1$ possibilities for each atom. For N identical atoms, $\Omega = (2J + 1)^N$, giving $S = k_B \ln(2J + 1)^N = Nk_B \ln(2J + 1)$.

The two steps of isentropic cooling (figure 5.6) are:

1. Isothermal magnetisation
   A field is applied to magnetise the material. The temperature of the sample is kept constant by a helium bath that removes heat from the sample. As seen from above, the entropy decreases on magnetisation.
2. Adiabatic demagnetisation
   The sample is now isolated (adiabatic) and the magnetic field is slowly reduced to zero (quasi-static). This demagnetisation increases the entropy of the magnetic moments. But because it is an adiabatic process ($Q = 0$), the entropic change is zero, thus there must be a compensating decrease in entropy elsewhere. This decrease in entropy takes place in the phonons, and hence the material cools.

5.4 Collective magnetism

Notice that we could understand the phenomena of diamagnetism and paramagnetism without assuming any explicit interaction among the magnetic moments (or among the electrons that gave rise to these magnetic moments).

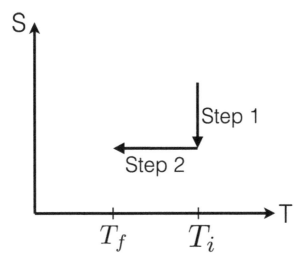

Figure 5.6. The entropy–temperature $S-T$ diagram showing the two steps of isentropic cooling.

For collective magnetism, there are interactions between the magnetic moments and these give rise to cooperative or collective phenomena (ferromagnetism, ferrimagnetism, and antiferromagnetism). These interactions are given the name 'exchange interactions'. The signature of cooperative phenomena is a spontaneous ordering of magnetic moments below a critical temperature. In ferro- and ferrimagnetism, this is called the Curie temperature T_C, and in antiferromagnetism, it is called the Néel temperature T_N.

We make a broad classification first:

- Insulators
  - have magnetic moments that are produced by partially filled electron shells;
  - can be ferro-, ferri-, or antiferromagnetic;
  - are well described by the Heisenberg model (see equation (5.163) later).
- Metals
  - Band magnetism (beyond the scope of this book):
    - ○ their conduction electrons are responsible for magnetism;
    - ○ their exchange interaction causes a spin-dependent band shift (for $T < T_C$) and thus a particular spin orientation is preferred;
    - ○ their simplest model is the Hubbard model.
  - Localised magnetism:
    - ○ has magnetic moments that are not produced by conduction electrons;
    - ○ has an exchange interaction that takes place between localised electrons that contribute magnetic moments and itinerant conduction electrons.

5.4.1 The direct exchange interaction

The physics behind the direct exchange interaction is extremely simple: Coulomb interaction and the Pauli exclusion principle.

We will illustrate this using the standard Heitler–London treatment in molecular physics. We will also 'derive' the Heisenberg model.

Consider the model of a hydrogen molecule (or two hydrogen atoms) (figure 5.7). The Hamiltonians of the two separate atoms (in position representation) are

$$\text{First atom: } H_0^a(1) = -\frac{\hbar^2}{2m_e}\nabla_{r_{a1}}^2 - \frac{e^2}{4\pi\varepsilon_0 r_{a1}} \tag{5.126}$$

$$\text{Second atom: } H_0^b(2) = -\frac{\hbar^2}{2m_e}\nabla_{r_{b2}}^2 - \frac{e^2}{4\pi\varepsilon_0 r_{b2}} \tag{5.127}$$

$$\text{Molecule: } H = H_0^a(1) + H_0^b(2) - \frac{e^2}{4\pi\varepsilon_0}\left(\frac{1}{r_{a2}} + \frac{1}{r_{b1}} - \frac{1}{r_{12}} - \frac{1}{R}\right) \tag{5.128}$$

The Hamiltonian of the molecule gives an unsolvable Schrödinger equation. Thus, we shall make an approximation using physically reasonable wave functions. Since there are two electrons involved, they have to obey the generalised Pauli exclusion principle known as 'exchange symmetry', which we will summarise now.

- The two-boson wave function obeys the following under exchange:

$$\Psi_{\text{boson}}(1, 2)\overset{1\leftrightarrow2}{\longrightarrow}\Psi_{\text{boson}}(2, 1) = \Psi_{\text{boson}}(1, 2). \tag{5.129}$$

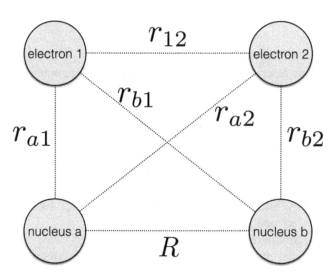

Figure 5.7. A hydrogen molecule consisting of two hydrogen atoms.

- The two-fermion wave function obeys the following under exchange:

$$\Psi_{\text{fermion}}(1, 2)\xrightarrow{1\leftrightarrow 2}\Psi_{\text{fermion}}(2, 1) = -\Psi_{\text{fermion}}(1, 2). \tag{5.130}$$

Now we make a guess at the spatial part of the two-electron wave function.

$$\Psi_{+} = \frac{1}{\sqrt{2}}(\psi^{a}(1)\psi^{b}(2) + \psi^{a}(2)\psi^{b}(1)) \tag{5.131}$$

$$\Psi_{-} = \frac{1}{\sqrt{2}}(\psi^{a}(1)\psi^{b}(2) - \psi^{a}(2)\psi^{b}(1)) \tag{5.132}$$

It appears that Ψ_{+} does not obey the two-fermion exchange symmetry, and indeed it does not, but we keep Ψ_{+} anyway.

For the spin part, we recall from quantum mechanics the results of adding the two spin-$\frac{1}{2}$ angular momenta:

$$s_1 = \frac{1}{2}, \ s_2 = \frac{1}{2} \implies s = 0, 1. \tag{5.133}$$

The (old) basis states are

$$\left| m_{s_1} = \frac{1}{2}, m_{s_2} = \frac{1}{2} \right\rangle = |\uparrow_1 \uparrow_2 \rangle \tag{5.134}$$

$$\left| m_{s_1} = \frac{1}{2}, m_{s_2} = -\frac{1}{2} \right\rangle = |\uparrow_1 \downarrow_2 \rangle \tag{5.135}$$

$$\left| m_{s_1} = -\frac{1}{2}, m_{s_2} = \frac{1}{2} \right\rangle = |\downarrow_1 \uparrow_2 \rangle \tag{5.136}$$

$$\left| m_{s_1} = -\frac{1}{2}, m_{s_2} = -\frac{1}{2} \right\rangle = |\downarrow_1 \downarrow_2 \rangle \tag{5.137}$$

and the new basis states in terms of the original basis are

$$|s = 0, m_s = 0\rangle = |0, 0\rangle = \frac{1}{\sqrt{2}}(|\uparrow_1 \downarrow_2 \rangle - |\downarrow_1 \uparrow_2 \rangle) = \chi_S(1, 2) \tag{5.138}$$

$$|s = 1, m_s = 1\rangle = |1, 1\rangle = |\uparrow_1 \uparrow_2 \rangle = \chi_T(1, 2) \tag{5.139}$$

$$|s = 1, m_s = 0\rangle = \frac{1}{\sqrt{2}}(|\uparrow_1 \downarrow_2 \rangle + |\downarrow_1 \uparrow_2 \rangle) = \chi_T(1, 2) \tag{5.140}$$

$$|s = 1, m_s = -1\rangle = |\downarrow_1 \downarrow_2 \rangle = \chi_T(1, 2), \tag{5.141}$$

where $\chi_S(1, 2)$ is the antisymmetric singlet state with $\chi_S(1, 2)\xrightarrow{1\leftrightarrow2}\chi_S(2, 1)=-\chi_S(1, 2)$. And $\chi_T(1, 2)$ are the symmetric triplet spin states with $\chi_T(1, 2)\xrightarrow{1\leftrightarrow2}\chi_T(2, 1) = \chi_T(1, 2)$.

We can now construct the full spatial × spin wave functions that obey two-fermion exchange symmetry. Note that

$$\text{'symmetric'} \times \text{'antisymmetric'} = \text{'antisymmetric'} \tag{5.142}$$

$$\text{'symmetric'} \times \text{'symmetric'} = \text{'symmetric'} \tag{5.143}$$

$$\text{'antisymmetric'} \times \text{'antisymmetric'} = \text{'symmetric'}. \tag{5.144}$$

The two types of antisymmetric wave functions can now be constructed:

$$\Psi_S = \Psi_+\chi_S \tag{5.145}$$

$$\Psi_S = \frac{1}{\sqrt{2}}(\psi^a(1)\psi^b(2) + \psi^a(2)\psi^b(1))\chi_S(1, 2) \tag{5.146}$$

$$\Psi_T = \Psi_-\chi_T \tag{5.147}$$

$$\Psi_T = \frac{1}{\sqrt{2}}(\psi^a(1)\psi^b(2) - \psi^a(2)\psi^b(1))\chi_T(1, 2). \tag{5.148}$$

Also, Ψ_S and Ψ_T obey the exchange symmetry, i.e. $\Psi_S(1, 2)\xrightarrow{1\leftrightarrow2}\Psi_S(2, 1) = -\Psi_S(1, 2)$ and $\Psi_T(1, 2)\xrightarrow{1\leftrightarrow2}\Psi_T(2, 1) = -\Psi_T(1, 2)$.

In the light of first-order perturbation theory, we need the energy expectation values, which are as follows (note that these are not eigenvalues!):

$$E_S = \int d1d2 \ \Psi_S^*H\Psi_S \tag{5.149}$$

$$E_T = \int d1d2 \ \Psi_T^*H\Psi_T. \tag{5.150}$$

Note that the difference between the two energies is,

$$E_S - E_T = \int d1d2 \ (\Psi_S^*H\Psi_S - \Psi_T^*H\Psi_T)$$
$$\mid \text{recall that } \chi_S \text{ and } \chi_T \text{ are orthogonal to each other} \tag{5.151}$$
$$\mid \text{and expand out the spatial parts}$$

$$E_S - E_T = \int d1d2 \ \psi^{a*}(1)\psi^{b*}(2)H\psi^a(2)\psi^b(1) \tag{5.152}$$

$$+ \int d1d2 \; \psi^{a*}(2)\psi^{b*}(1)H\psi^{a}(1)\psi^{b}(2) \tag{5.153}$$

$$E_S - E_T \equiv \mathscr{J} \tag{5.154}$$

which is called the 'exchange integral'.

We now want to make an effective Hamiltonian that treats the wave functions Ψ_S and Ψ_T as eigenstates and E_S and E_T as the respective eigenvalues.

To do so, we first need to note this manipulation:

$$\text{Consider } (\vec{S}_1 + \vec{S}_2)^2 = \vec{S}_1^2 + \vec{S}_2^2 + 2\vec{S}_1 \cdot \vec{S}_2 \tag{5.155}$$

$$\vec{S}_1 \cdot \vec{S}_2 = \frac{1}{2}\left[(\vec{S}_1 + \vec{S}_2)^2 - \vec{S}_1^2 - \vec{S}_2^2 \right] \tag{5.156}$$

$$| \text{ note that } \vec{S}_1 + \vec{S}_2 = \vec{S}$$

$$\vec{S}_1 \cdot \vec{S}_2 = \frac{1}{2}\left[\vec{S}^2 - \vec{S}_1^2 - \vec{S}_2^2 \right] \tag{5.157}$$

$$| \text{ recall that } \vec{S}^2 = s(s + 1) \text{ (recall that they are dimensionless)}$$

$$\vec{S}_1 \cdot \vec{S}_2 = \frac{1}{2}[s(s + 1) - s_1(s_1 + 1) - s_2(s_2 + 1)] \tag{5.158}$$

$$| \text{ and for } s_1 = \frac{1}{2}, \; s_2 = \frac{1}{2} \text{ we have } s = 0 \text{ and } s = 1$$

$$\vec{S}_1 \cdot \vec{S}_2 = \frac{1}{2}\left[s(s + 1) - \frac{3}{2} \right] \tag{5.159}$$

$$\vec{S}_1 \cdot \vec{S}_2 = \begin{cases} -\dfrac{3}{4} & \text{for } s = 0 \\[2mm] \dfrac{1}{4} & \text{for } s = 1 \end{cases}. \tag{5.160}$$

Thus, if we define the 'effective' Hamiltonian,

$$H_{\text{eff}} = \frac{1}{4}(E_S + 3E_T) - (E_S - E_T)\vec{S}_1 \cdot \vec{S}_2 \tag{5.161}$$

it has two eigenvalues,

$$\text{for } s = 0, \quad H_{\text{eff}}|s = 0\rangle = \left(\frac{1}{4}(E_S + 3E_T) - (E_S - E_T)\left(-\frac{3}{4}\right) \right)|s = 0\rangle = E_S|s = 0\rangle$$

$$\text{for } s = 1, \quad H_{\text{eff}}|s = 1\rangle = \left(\frac{1}{4}(E_S + 3E_T) - (E_S - E_T)\left(\frac{1}{4}\right) \right)|s = 1\rangle = E_T|s = 1\rangle.$$

Using the notation of the exchange integral, $\mathscr{J}$,

$$H_{\text{eff}} = \frac{1}{4}(E_S + 3E_T) - \mathscr{J}\vec{S}_1 \cdot \vec{S}_2. \tag{5.162}$$

The first term is a constant and can always be absorbed by a shift. This gives the Heisenberg Hamiltonian for a two-spin system,

$$H_{\text{Heisenberg}} \equiv -\mathscr{J}\vec{S}_1 \cdot \vec{S}_2. \tag{5.163}$$

A few comments are in order:

- What we have done is to show that the Coulomb interaction (not explicitly shown here) and the Pauli exclusion principle (via exchange symmetry) can give rise to singlet spin states (that prefer opposite spins) and triplet spin states (that prefer parallel spins) that have different energy values. The singlet state turns out to be lower in energy; thus, in this case, the opposite spin arrangement is preferred.
- This is not really a derivation of the Heisenberg Hamiltonian. This is just a justification that even without the presence of a magnetic field, other interactions can still bring about a preferred spin arrangement.
- For nearest-neighbour interaction in a multispin system, we can write the Heisenberg Hamiltonian as

$$H_{\text{Heisenberg}} = -\sum_{\langle i,j \rangle} \mathscr{J}_{ij}\vec{S}_i \cdot \vec{S}_j \tag{5.164}$$

where $\langle i, j \rangle$ means 'over nearest neighbours' and the exchange integral $\mathscr{J}_{ij}$ is usually taken to be a parameter.

- There are different variants of such spin–spin interaction models; they are compared below:

The Heisenberg model: $\quad H = -\sum_{\langle i,j \rangle} \mathscr{J}_{ij}\left(S_i^x S_j^x + S_i^y S_j^y + S_i^z S_j^z\right) \tag{5.165}$

The Ising model: $\quad H = -\sum_{\langle i,j \rangle} \mathscr{J}_{ij} S_i^z S_j^z \tag{5.166}$

The XY model: $\quad H = -\sum_{\langle i,j \rangle} \mathscr{J}_{ij}\left(S_i^x S_j^x + S_i^y S_j^y\right) \tag{5.167}$

- The other types of exchange interactions are called 'indirect exchange interactions'. The different classes of indirect exchange interactions are:
  - Superexchange interaction
    This exchange interaction is mediated by a non-magnetic ion in between the magnetic ions.

- RKKY exchange interaction
    This was devised by Ruderman, Kittel, Kasuya, and Yosida. Exchange is mediated by valence electrons and thus the exchange integral is distance dependent.
- Double exchange
    Magnetic ions can have different oxidation states and electron hopping between ions results in an extended ferromagnetic (i.e. parallel spin) exchange interaction.

5.4.2 Ferromagnetic order

Ferromagnetism refers to spontaneous magnetisation without an applied field, such that the magnetic moments all point in the same direction. Beyond the so-called Curie temperature T_C, the interaction breaks down and the material becomes paramagnetic.

There are two aspects of ferromagnetism:

1. Within a domain, the magnetic moments are all aligned.
2. Domains interact to produce other magnetic effects, such as hysteresis.

We will discuss the first aspect: interactions within a domain.

5.4.2.1 The Weiss mean field theory of ferromagnetism (the simplest theory of ferromagnetism)

A theory of ferromagnetism (and collective magnetism) must be able to at least show a phase transition between paramagnetism and ferromagnetism. The transition temperature should be calculable from the theory.

To create the mean field theory, Weiss took the paramagnetic Hamiltonian $H^{para} = g_{eff}\mu_B \vec{B} \cdot \vec{J}$ and wrote $\vec{B} \longrightarrow \vec{B} + \vec{B}^{MF}$, where $\vec{B}$ is the external field and $\vec{B}^{MF}$ is some kind of internal mean field (or exchange interaction or molecular field).

He further assumed that the mean field is proportional to the magnetisation:

$$\vec{B}^{MF} = \lambda \vec{M}, \tag{5.168}$$

where λ is a parameter and $\vec{M}$ is the magnetisation. The Hamiltonian for the theory is then

$$H^{Weiss} = \sum_{i=1}^{z} g_{eff}\mu_B \vec{J}_i \cdot (\vec{B} + \vec{B}^{MF}) = \sum_{i=1}^{z} g_{eff}\mu_B \vec{J}_i \cdot (\vec{B} + \lambda \vec{M}). \tag{5.169}$$

Since this Hamiltonian is so similar to the one used for Curie paramagnetism (equation (5.34)), we can immediately carry the results over. We carry over the magnetisation expression

$$M = g_{eff}\mu_B J B_J(xJ) \tag{5.170}$$

and $B_J(xJ)$ is the Brillouin function with the modification

$$x = \frac{g_{eff}\mu_B(B + \lambda M)}{k_B T}. \tag{5.171}$$

For an external field $B = 0$,

$$M = g_{\text{eff}}\mu_B J B_J\left(\frac{g_{\text{eff}}\mu_B\lambda M}{k_B T}J\right). \tag{5.172}$$

Thus, solving for M is not trivial but can be done graphically. Let the equations be written as two equations

$$y = M \tag{5.173}$$

$$y = g_{\text{eff}}\mu_B J B_J\left(\frac{g_{\text{eff}}\mu_B\lambda M}{k_B T}J\right) \tag{5.174}$$

The graphical solution of M requires us to find the intersection points of these two curves.

$M = 0$ is always a solution but it is trivial. We shall now discuss the nontrivial solutions. We see that for different temperature values, $y = g_{\text{eff}}\mu_B J B_J(...)$ can change, and we have three possible cases (figure 5.8):

1. For case (a):
   The only solution is for $M = 0$, which is the paramagnetism situation (recall that there is no external field). So it represents $T > T_C$.
2. For case (c):

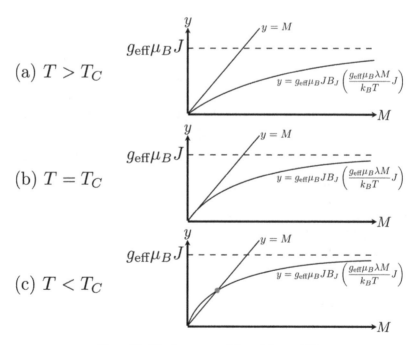

Figure 5.8. The three cases of the solutions of M.

The nontrivial solution is marked by the dot. This nonzero M represents spontaneous magnetisation (ferromagnetism), which occurs for $T < T_C$.

3. For case (b):

The second intersection point is 'about to appear'. Thus, it makes sense to define this temperature as the transition temperature T_C. We can actually use this definition to solve for T_C. The condition is that for small M, the gradient of $y = g_{\text{eff}}\mu_B J B_J(\dots)$ should be one, therefore (write $T = T_C$),

$$\frac{d}{dM} g_{\text{eff}}\mu_B J B_J(\dots)\bigg|_{M=0} = 1 \tag{5.175}$$

$$\frac{d}{dM} g_{\text{eff}}\mu_B J B_J\left(\frac{g_{\text{eff}}\mu_B \lambda M}{k_B T_C} J\right)\bigg|_{M=0} = 1 \tag{5.176}$$

since M is small, we expand $B_J(xJ) \approx \dfrac{J+1}{3} x \;|$

$$g_{\text{eff}}\mu_B J \frac{d}{dM}\left(\frac{J+1}{3}\frac{g_{\text{eff}}\mu_B \lambda M}{k_B T_C}\right) = 1 \tag{5.177}$$

$$\frac{g_{\text{eff}}^2 \mu_B^2 \lambda J(J+1)}{3k_B T_C} = 1 \tag{5.178}$$

$$T_C = \frac{g_{\text{eff}}^2 \mu_B^2 \lambda J(J+1)}{3k_B}. \tag{5.179}$$

We will now work out the behaviour of M near T_C. We recall the expression for magnetisation M:

$$M = g_{\text{eff}}\mu_B J B_J\left(\frac{g_{\text{eff}}\mu_B \lambda M}{k_B T} J\right) \tag{5.180}$$

$\mid$ substitute $\dfrac{g_{\text{eff}}\mu_B \lambda J}{k_B} = \dfrac{3T_C}{g_{\text{eff}}\mu_B(J+1)}$

$$M = g_{\text{eff}}\mu_B J B_J\left(\frac{M}{T}\frac{3T_C}{g_{\text{eff}}\mu_B(J+1)}\right)$$

$\mid$ since M is small near T_C, we can expand B_J as a Taylor series

$\mid$ this time we need to retain one more order: $\coth y = \dfrac{1}{y} + \dfrac{y}{3} - \dfrac{y^3}{45} + \cdots$ (5.181)

$\mid$ we already know the first two terms give $B_J(xJ) \approx \dfrac{J+1}{3} x$

$$M = g_{\text{eff}}\mu_B J \left[\frac{J+1}{3J} \frac{M}{T} \frac{3T_C}{g_{\text{eff}}\mu_B(J+1)} - \frac{2J+1}{2J} \left(\frac{2J+1}{2J} \right)^3 \frac{1}{45} \left(\frac{M}{T} \frac{3T_C}{g_{\text{eff}}\mu_B(J+1)} \right)^3 \right.$$
$$\left. + \frac{1}{2J} \left(\frac{1}{2J} \right)^3 \frac{1}{45} \left(\frac{M}{T} \frac{3T_C}{g_{\text{eff}}\mu_B(J+1)} \right)^3 \right] \tag{5.182}$$

| remove one common factor of M

$$1 = \frac{T_C}{T} \left[1 + \frac{1-(2J+1)^4}{45(2J)^4} \left(\frac{T_C}{T} \right)^2 \frac{27M^2 J}{g_{\text{eff}}^2 \mu_B^2 (J+1)^3} \right] \tag{5.183}$$

$$\frac{T}{T_C} - 1 = \frac{1-(2J+1)^4}{45(2J)^4} \frac{27J}{g_{\text{eff}}^2 \mu_B^2 (J+1)^3} \left(\frac{T_C}{T} \right)^2 M^2 \tag{5.184}$$

$$\frac{T}{T_C} - 1 = -\frac{3(2J^2+2J+1)}{10g_{\text{eff}}^2 \mu_B^2 J^2(J+1)^2} \left(\frac{T_C}{T} \right)^2 M^2 \tag{5.185}$$

$$M^2 = \frac{10g_{\text{eff}}^2 \mu_B^2 J^2(J+1)^2}{3(2J^2+2J+1)} \left(1 - \frac{T}{T_C} \right) \left(\frac{T}{T_C} \right)^2 \tag{5.186}$$

$$M^2 \propto \left(\frac{T}{T_C} \right)^2 \left(1 - \frac{T}{T_C} \right) \tag{5.187}$$

| for $T \to T_C$, we can set $\left(\dfrac{T}{T_C} \right)^2 \to 1$

$$M \propto \sqrt{1 - \frac{T}{T_C}} . \tag{5.188}$$

This relationship is illustrated in figure 5.9.

Finally, we can discuss the susceptibility of the ferromagnet in Weiss mean field theory. We need to have a (small) magnetic field in the system. Recall the magnetisation expression with an external field:

$$M = g_{\text{eff}}\mu_B J B_J(xJ)$$

| with modification $x = \dfrac{g_{\text{eff}}\mu_B(B+\lambda M)}{k_B T}$ and write $\dfrac{g_{\text{eff}}\mu_B}{k_B} = \dfrac{3T_C}{g_{\text{eff}}\mu_B \lambda J(J+1)}$ $\qquad$ (5.189)

$$M = g_{\text{eff}}\mu_B J B_J \left(\frac{3T_C}{g_{\text{eff}}\mu_B \lambda J(J+1)} \frac{B+\lambda M}{T} J \right) \tag{5.190}$$

| at $T > T_C$, the external field causes a small magnetization

| thus recall the expansion $B_J(xJ) \approx \dfrac{J+1}{3} x = \dfrac{J+1}{3J} xJ$

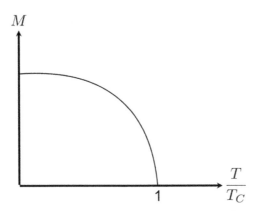

Figure 5.9. The relationship between magnetisation M and T/T_C as given by Weiss mean field theory.

$$M \approx g_{\text{eff}}\mu_B J\left(\frac{J+1}{3J}\frac{3T_C}{g_{\text{eff}}\mu_B\lambda J(J+1)}\frac{B+\lambda M}{T}J\right) \qquad (5.191)$$

$$M = \frac{T_C}{\lambda}\frac{B+\lambda M}{T} \qquad (5.192)$$

$$M = \frac{\dfrac{T_C}{T}B}{\lambda\left(1-\dfrac{T_C}{T}\right)} \qquad (5.193)$$

$$\chi = \frac{\mu_0 M}{B} \implies \chi = \frac{\mu_0\dfrac{T_C}{T}}{\lambda\left(1-\dfrac{T_C}{T}\right)} \qquad (5.194)$$

$$\chi = \frac{\mu_0 T_C}{\lambda}\frac{1}{T-T_C}. \qquad (5.195)$$

The expression $\chi \propto \frac{1}{T-T_C}$ is the famous Curie–Weiss Law. Do not confuse this with Curie's Law of paramagnetism.

A few comments are now in order:

- Recall that Weiss mean field theory is based on the assumption that there is an internal field that creates interactions between the magnetic moments. Weiss had no idea of the exchange interaction! Thus, it is already remarkable that this theory contains the phase transition between paramagnetism and ferromagnetism. Measuring T_C gives a fit for λ.

- Further analysis shows the failures of Weiss theory:
  - For $T \to 0$, the expression of magnetisation from Weiss theory and that from experiments do not agree.
  - The $T \to T_C$ expression obtained earlier: $M \propto \sqrt{1 - \dfrac{T}{T_C}}$ also does not agree with the experimental expression: $M \propto \left(1 - \dfrac{T}{T_C}\right)^{1/3}$.
  - The specific heat for $T > T_C$ predicted by Weiss theory is zero, which is completely incorrect. Experimental results seem to indicate a logarithmic singularity as $T \to T_C$ for the specific heat. Weiss theory could not even be used to provide an explanation for that.

- Weiss mean field theory can be 'derived' from the Heisenberg Hamiltonian by the following argument: consider the Heisenberg Hamiltonian with nearest neighbour (NN) and next nearest neighbour (NNN) interactions, $H_{\text{Heisenberg}} = -\mathscr{J} \sum_i \vec{S}_i \cdot \vec{S}_{i+1} - \mathscr{J}' \sum_i \vec{S}_i \cdot \vec{S}_{i+2}$. We then take the average of the NN and NNN spins to get $H_{\text{Heisenberg}} \approx -\mathscr{J} \sum_i \vec{S}_i \cdot \langle \vec{S}_{i+1} \rangle - \mathscr{J}' \sum_i \vec{S}_i \cdot \langle \vec{S}_{i+2} \rangle$, and the average spin $\langle \vec{S} \rangle$ is proportional to the magnetisation. Thus, $H_{\text{Heisenberg}}$ and H^{Weiss} have similar forms.

5.4.2.2 Magnons (quantised spin waves)

At low temperatures, a low-energy excitation called a spin wave propagates among the interacting spins. Quantised spin waves are called magnons. They are extremely similar to phonons, which are quantised lattice waves.

We avoid any semiclassical derivations, as this is really a quantum mechanical effect. It can be described by the Heisenberg model and we shall give a simplified derivation of spin waves.

We consider a 1D chain of N interacting spins.[3] Thus, each spin has two NNs except for the first and last spin. The Heisenberg Hamiltonian can be written as

$$H_{\text{Heisenberg}} = - \sum_{i=1}^{N-1} \mathscr{J}_{i,i+1} \vec{S}_i \cdot \vec{S}_{i+1} \tag{5.196}$$

$$\mid \text{take } \mathscr{J}_{i,i+1} = \mathscr{J}_{i+1,i} = \mathscr{J}$$

$$H_{\text{Heisenberg}} = -\mathscr{J} \sum_{i=1}^{N-1} \vec{S}_i \cdot \vec{S}_{i+1} \tag{5.197}$$

---

[3] Strictly speaking, we require translational invariance in this derivation. We can achieve this by either taking $N \to \infty$ or by joining the 'head' and 'tail' of the spin chain together (a periodic boundary condition). We neglect this detail in this derivation and an improved derivation will be done in the exercises.

$$H_{\text{Heisenberg}} = -\mathcal{J} \sum_{i=1}^{N-1} S_i^z S_{i+1}^z + S_i^x S_{i+1}^x + S_i^y S_{i+1}^y$$

| write S^x and S^y in terms of raising and lowering operators   (5.198)

| thus recall $S_i^x = \dfrac{1}{2}\left(S_i^+ + S_i^-\right)$ and $S_i^y = \dfrac{1}{2i}\left(S_i^+ - S_i^-\right)$

$$H_{\text{Heisenberg}} = -\mathcal{J} \sum_{i=1}^{N-1} S_i^z S_{i+1}^z + \frac{1}{2}(S_i^+ + S_i^-)\frac{1}{2}(S_{i+1}^+ + S_{i+1}^-) + \frac{1}{2i}(S_i^+ - S_i^-)\frac{1}{2i}(S_{i+1}^+ - S_{i+1}^-)$$
$$ \tag{5.199}$$
$$H_{\text{Heisenberg}} = -\mathcal{J} \sum_{i=1}^{N-1} S_i^z S_{i+1}^z + \frac{1}{2}(S_i^+ S_{i+1}^- + S_i^- S_{i+1}^+)$$

We now specialise this to spin-$\frac{1}{2}$ for a simplified discussion; thus, there are only two states: $m_s = \frac{1}{2}$ up and $m_s = -\frac{1}{2}$ down. The $T = 0$ ground state of the ferromagnet can be written as

$$|0\rangle = |\uparrow_1 \cdots \uparrow_N \rangle. \tag{5.200}$$

The lowest excited state is one spin flip at site j.

$$|1_j\rangle = |\uparrow_1 \cdots \uparrow_{j-1} \downarrow_j \uparrow_{j+1} \cdots \uparrow_N \rangle \tag{5.201}$$

The total spin of the system changed by one (hence the notation). Thus, this is a bosonic excitation. We want to find the energy eigenvalue of such one-spin flip states. First, consider

$$H_{\text{Heisenberg}}|0\rangle = \left(-\mathcal{J} \sum_{i=1}^{N-1} S_i^z S_{i+1}^z + \frac{1}{2}(S_i^+ S_{i+1}^- + S_i^- S_{i+1}^+)\right)|\uparrow_1 \cdots \uparrow_N \rangle \tag{5.202}$$

| note that $S_i^+|\uparrow_i \rangle = 0$ and $S_i^z|\uparrow_i \rangle = \dfrac{1}{2}|\uparrow_i \rangle$

$$H_{\text{Heisenberg}}|0\rangle = -\mathcal{J} \sum_{i=1}^{N-1} S_i^z S_{i+1}^z|\uparrow_1 \cdots \uparrow_N \rangle \tag{5.203}$$

$$H_{\text{Heisenberg}}|0\rangle = -\mathcal{J}\frac{N-1}{4}|0\rangle, \tag{5.204}$$

and also

$$H_{\text{Heisenberg}}|1_j\rangle = \left(-\mathcal{J} \sum_{i=1}^{N-1} S_i^z S_{i+1}^z + \frac{1}{2}(S_i^+ S_{i+1}^- + S_i^- S_{i+1}^+)\right)|\uparrow_1 \cdots \downarrow_j \cdots \uparrow_N \rangle \tag{5.205}$$

| note the only relation we need is $S_j^+|\downarrow_j \rangle = |\uparrow_j \rangle$

$$H_{\text{Heisenberg}}|1_j\rangle = -\mathscr{J}\sum_{i=1}^{N-1}\left(S_i^z S_{i+1}^z - \frac{1}{2}(|\uparrow_1\cdots\uparrow_j\downarrow_{j+1}\cdots\uparrow_N\rangle + |\uparrow_1\cdots\downarrow_{j-1}\uparrow_j\cdots\uparrow_N\rangle)\right)$$

$$H_{\text{Heisenberg}}|1_j\rangle = -\mathscr{J}\left((S_1^z S_2^z + \cdots + S_{j-1}^z S_j^z + S_j^z S_{j+1}^z + \cdots + S_{N-1}^z S_N^z)|1_j\rangle\right.$$
$$\left. + \frac{1}{2}(|1_{j+1}\rangle + |1_{j-1}\rangle)\right) \tag{5.206}$$

$$|\text{ note that } S_j^z|\uparrow_1\cdots\downarrow_j\cdots\uparrow_N\rangle = -\frac{1}{2}|\uparrow_1\cdots\downarrow_j\cdots\uparrow_N\rangle$$

$$H_{\text{Heisenberg}}|1_j\rangle = -\mathscr{J}\left(\frac{N-3}{4} - \frac{1}{4} - \frac{1}{4}\right)|1_j\rangle - \frac{1}{2}\mathscr{J}(|1_{j+1}\rangle + |1_{j-1}\rangle) \tag{5.207}$$

$$H_{\text{Heisenberg}}|1_j\rangle = -\mathscr{J}\frac{N-5}{4}|1_j\rangle - \frac{1}{2}\mathscr{J}\left(|1_{j+1}\rangle + |1_{j-1}\rangle\right). \tag{5.208}$$

Now consider the plane wave state $|q\rangle = \frac{1}{\sqrt{N}}\sum_j e^{iqR_j}|1_j\rangle$ (this is inspired by phonons)

$$H_{\text{Heisenberg}}|q\rangle = H_{\text{Heisenberg}}\frac{1}{\sqrt{N}}\sum_j e^{iqR_j}|1_j\rangle$$

$$|\text{ where } R_j \text{ is the distance from the first spin to the } j\text{th spin} \tag{5.209}$$
$$|\text{ let } a \text{ be the spacing between adjacent spins, so } R_j = ja$$

$$H_{\text{Heisenberg}}|q\rangle = \frac{1}{\sqrt{N}}\sum_j e^{iqR_j}\left(-\mathscr{J}\frac{N-5}{4}|1_j\rangle - \frac{1}{2}\mathscr{J}(|1_{j+1}\rangle + |1_{j-1}\rangle)\right) \tag{5.210}$$

$$H_{\text{Heisenberg}}|q\rangle = \frac{1}{\sqrt{N}}\sum_j\left(-\mathscr{J}\frac{N-5}{4}e^{iqR_j}|1_j\rangle - \frac{1}{2}\mathscr{J}\left(e^{iqR_j}e^{iqa}e^{-iqa}|1_{j+1}\rangle\right.\right.$$
$$\left.\left. + e^{iqR_j}e^{-iqa}e^{iqa}|1_{j-1}\rangle\right)\right) \tag{5.211}$$

$$|\quad \text{note } R_j + a = R_{j+1},\ R_j - a = R_{j-1} \text{ and } \frac{1}{\sqrt{N}}\sum_j e^{iqR_{j\pm1}}|1_{j\pm1}\rangle = |q\rangle$$
$$|\text{ the last relation requires translational invariance, which we ignore here}$$

$$H_{\text{Heisenberg}}|q\rangle = \left(-\mathscr{J}\frac{N-5}{4} - \frac{1}{2}\mathscr{J}(e^{-iqa} + e^{iqa})\right)|q\rangle \tag{5.212}$$

$$H_{\text{Heisenberg}}|q\rangle = \left(-\mathscr{J}\frac{N-1}{4} + \mathscr{J}(1 - \cos(qa))\right)|q\rangle \tag{5.213}$$

Thus, $|q\rangle$ is an eigenstate of $H_{\text{Heisenberg}}$. The state $|q\rangle$ is a sum of all one-spin flip states $|1_j\rangle$ with phase factors e^{iqR_j}; thus, $|q\rangle$ is a 'plane wave' of one-spin flips (figure 5.10).

If we shift up the Heisenberg Hamiltonian by $\mathscr{J}\frac{N-1}{4}$, we have

$$H'_{\text{Heisenberg}} = H_{\text{Heisenberg}} + \mathscr{J}\frac{N-1}{4} \tag{5.214}$$

$$\text{then, } H'_{\text{Heisenberg}}|0\rangle = 0|0\rangle \tag{5.215}$$

$$H'_{\text{Heisenberg}}|q\rangle = \mathscr{J}\big(1 - \cos(qa)\big)|q\rangle. \tag{5.216}$$

Thus the dispersion of a 'plane wave' of one-spin flips or a spin wave is $E(q) = \mathscr{J}(1 - \cos(qa))$ (see figures 5.10 and 5.11).

We will not carry on to quantise the spin waves into magnons, as this would involve second quantisation notation and Holstein–Primakoff transformation, which are beyond the scope of this book.

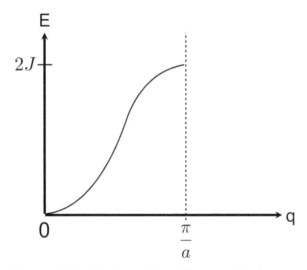

Figure 5.10. The dispersion relation of ferromagnetic spin waves.

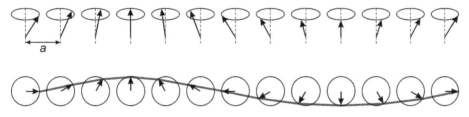

Figure 5.11. The semiclassical picture of spin waves. Note that it is just a semiclassical picture. Reproduced from [1] John Wiley & Sons.

5.4.2.3 Ferromagnetic domains

[Why do domains form?] Four types of energy compete in ferromagnets. The result is that ferromagnets are not made of a single domain of aligned spins but domains of spins aligned in different directions. Such a structure of domains actually results in a lower energy for the material.

1. Exchange energy:

    This short-ranged interaction was modelled by the Heisenberg Hamiltonian $H = -\mathcal{J}\vec{S}_i \cdot \vec{S}_j$, and parallel alignment lowers the energy.

2. Magnetostatic energy:

    The simplest way to describe this long-ranged energy is to consider that this energy is used to 'maintain' the external magnetic field of the magnetised material. If we break the material, making two dipoles, and anti-align them, the external field (and thus the energy) will be greatly reduced.[4] This favours anti-alignment and the formation of domains with anti-aligned moments (figure 5.12).

3. Anisotropy energy:

    If we only consider the first two types of energies, then the most favourable condition is that the magnetic moments gradually flip over from one end of the material to the other end, i.e. the width of the domain wall is the width of the material! Due to the crystalline structure of the material, certain directions require less energy for the magnetic moments to align with (easy axes). The hard axes are defined oppositely. The width of the domain would be a balance of exchange and anisotropy energies, as most of the rotated spins would not be along the easy axis, and this costs anisotropic energy.

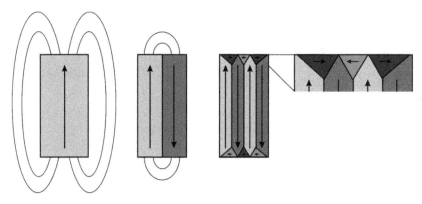

Figure 5.12. The formation of domains (from left to right) resulting in a reduction of magnetostatic energy. The rightmost diagram shows closure domains. Reprinted by permission from Springer Nature [2], Copyright (2007).

---

[4] One can also think of the fact that a broken magnet also tends to anti-align.

4. Magnetoelastic energy:

Changes in magnetisation result in (small) elastic distortions of the material. The larger the domains, the more elastic energy is needed. Thus, the formation of smaller domains (such as closure domains) is favoured by magnetoelastic energy.

[Types of Domain Walls]

- Bloch wall: magnetisation rotates in a plane parallel to the plane of the domain wall. See figure 5.13(a).
- Néel wall: magnetisation rotates in a plane perpendicular to the plane of the domain wall. See figure 5.13(b).
- There is a 90° or 180° change in magnetic moments between adjacent domains. See figure 5.14.

[Estimation of the energy of a 180° Bloch wall] Assume that a π change in moments occurs over N sites; thus, each spin is rotated by $\frac{\pi}{N}$ radians with respect to

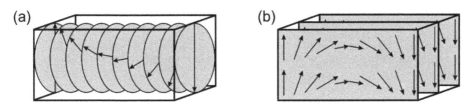

Figure 5.13. (a) the Bloch wall and (b) the Néel wall. Reprinted by permission from Springer Nature [2], Copyright (2007).

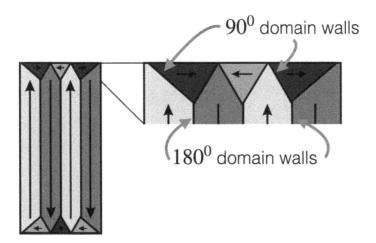

Figure 5.14. A 180^0 domain wall and a 90^0 domain wall. Adapted with permission from Springer Nature [2], Copyright (2007).

its neighbour. The exchange energy E_{ex} is estimated using the Heisenberg Hamiltonian.

$$H_{\text{Heisenberg}} = -\mathscr{J}\sum_i \vec{S}_i \cdot \vec{S}_{i+1} \qquad (5.217)$$

'semi–classically' speaking $\implies$ $E = -\mathscr{J}\sum_i S_i S_{i+1} \cos\theta \qquad (5.218)$

$$E = -\mathscr{J}NS^2 \cos\left(\frac{\pi}{N}\right) \qquad (5.219)$$

| assume the angle is small and expand $\cos x \approx 1 - \dfrac{x^2}{2}$

$$E = -\mathscr{J}NS^2 + \mathscr{J}NS^2\left(\frac{\pi}{N}\right)^2 \times \frac{1}{2} \qquad (5.220)$$

The first term is for parallel alignment. The exchange energy is thus the second term,

$$E_{ex} \approx \frac{1}{2}\mathscr{J}NS^2\left(\frac{\pi}{N}\right)^2. \qquad (5.221)$$

For the anisotropy energy density, we assume a simple form:

$$E = K\sin^2\theta \quad K > 0, \qquad (5.222)$$

where K is an anisotropy constant. This model means that each spin prefers $\theta = 0$ or $\theta = \pi$. Thus, the total anisotropy energy is

$$E_{\text{ani}} = \sum_{i=1}^{N} K\sin^2\theta_i \qquad (5.223)$$

| assume a large number of spins and take thecontinuum limit

$$E_{\text{ani}} = \frac{N}{\pi}\int_0^\pi K\sin^2\theta d\theta \qquad (5.224)$$

| the 'volume' correction factor $\dfrac{N}{\pi}$ is needed since $\Delta\theta = \dfrac{\pi}{N}$

$$E_{\text{ani}} = \frac{N}{\pi}K\int_0^\pi \left(\frac{1}{2} - \frac{1}{2}\cos 2\theta\right)d\theta \qquad (5.225)$$

$$E_{\text{ani}} = \frac{N}{\pi}K\frac{\pi}{2} \qquad (5.226)$$

$$E_{\text{ani}} = \frac{NK}{2}. \tag{5.227}$$

The total energy in the domain wall is

$$E_{\text{total}} = E_{ex} + E_{\text{ani}} = \frac{\mathscr{J}S^2\pi^2}{2N} + \frac{NK}{2}. \tag{5.228}$$

Thus, this expression is in line with the earlier discussion that exchange energy favours a wide domain wall N. We minimise E_{total} with respect to N:

$$\frac{dE_{\text{total}}}{dN} = 0 \implies -\frac{\mathscr{J}S^2\pi^2}{2N^2} + \frac{K}{2} = 0 \tag{5.229}$$

$$N = \pi S \sqrt{\frac{\mathscr{J}}{K}}. \tag{5.230}$$

Thus, the width of the domain wall depends on the ratio between the exchange constant and the anisotropy constant.

[Magnetisation and Hysteresis] The most obvious consequence is that the Curie temperature of iron $T_C \approx 1000K$, so why is iron not magnetised at room temperature? The answer is that each domain is magnetised in one direction, and these directions are quite random, so there is no net magnetisation.

The other consequence is that soft magnetic materials only require a small magnetic field (as low as 10^{-6} T) to reach saturation magnetisation. The short version of the explanation is that since there is already alignment within a domain, only a small magnetic field is needed to align the domains.

We now consider the process of magnetisation of a ferromagnet with domains (figure 5.15).

1. We assume there is no net magnetisation in the material to begin with. The magnetisation M increases with the applied field H. There is a linear region where the process is reversible: removing the field results in no magnetisation. Increasing the field beyond the linear region brings the magnetisation to a maximum saturated value M_s. Microscopically, domains aligned parallel to the field grow in size, and this is reversible for small domain wall motion. Saturated magnetisation is reached when domains that are aligned to H have reached their largest size.

2. From M_s, when the external field is decreased to zero, there is a remanent magnetisation M_r. Microscopically, domains that are aligned to H decrease in size when H is decreased. However, domain walls may get pinned by strains and impurities, resulting in net magnetisation even when $H = 0$.

3. An opposite field is applied to 'kill' M_r. The amount of opposite field is denoted by H_c, also known as the coercive field. Microscopically, H_c forces the pinned domain walls to move so that net magnetisation can reach zero. The size of H_c thus depends on the material preparation and properties.

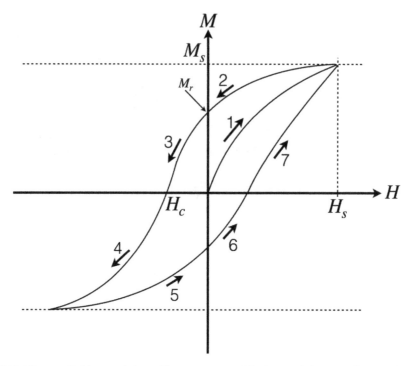

Figure 5.15. The so-called hysteresis loop. The seven parts of the hysteresis loop are discussed in the text. M_s—saturation magnetisation, M_r—remanent magnetisation, and H_c—coercive field.

4. A large enough opposite field causes an opposite saturation magnetisation.
5. This is the repeat of process 2.
6. This is the repeat of process 3.
7. This is the repeat of process 4.

5.4.3 Antiferromagnetic and ferrimagnetic orders

Antiferromagnetism and ferrimagnetism are simply cases in which adjacent spins prefer to anti-align with each other. In antiferromagnetism, the antiparallel moments are of the same magnitude and they cancel to give no net magnetisation. In ferrimagnetism, the antiparallel moments are unequal in magnitude, and this results in a net magnetisation.

We will discuss Weiss mean field theory for antiferromagnetism. Weiss mean field theory for ferrimagnetism is left as an exercise.

We should also note that the Heisenberg model can be modified to describe antiferromagnetism and ferrimagnetism. To favour anti-alignment, the exchange energy simply needs to be negative.

$$H_{\text{Heisenberg}} = -\mathscr{J} \sum_{i=1} \vec{S}_i \cdot \vec{S}_{i+1} = |\mathscr{J}| \sum_{i=1} \vec{S}_i \cdot \vec{S}_{i+1} \quad \mathscr{J} < 0 \tag{5.231}$$

5.4.3.1 The Weiss mean field theory of antiferromagnetism

We assume only NN interactions, so we have two kinds of effective fields:

$$B_\uparrow = -|\lambda|M_\downarrow + B \quad \text{and} \quad B_\downarrow = -|\lambda|M_\uparrow + B, \tag{5.232}$$

where:

- The term $-|\lambda|M_\downarrow$ is the mean field as seen by the up-spins. Since we assume NN interactions, up-spins 'see' down-spin magnetisation. The opposite is true for the term $-|\lambda|M_\uparrow$. We have also assumed that there is only one mean field constant λ. λ is negative because the exchange interaction is negative.
- The term B is the external field.

The Hamiltonian is then

$$H^{\text{Weiss}} = \sum_{i=1}^{z} g_{\text{eff}}\mu_B \vec{J}_i \cdot \vec{B}_\uparrow + \sum_{i=1}^{z} g_{\text{eff}}\mu_B \vec{J}_i \cdot \vec{B}_\downarrow. \tag{5.233}$$

Thus, we can again carry over the Curie paramagnetism expression, (set $B = 0$):

$$M_\uparrow = g_{\text{eff}}\mu_B J B_J\left(-\frac{g_{\text{eff}}\mu_B|\lambda|M_\downarrow}{k_B T}J\right) \tag{5.234}$$

$$M_\downarrow = g_{\text{eff}}\mu_B J B_J\left(-\frac{g_{\text{eff}}\mu_B|\lambda|M_\uparrow}{k_B T}J\right). \tag{5.235}$$

But in an antiferromagnet below T_C (and $B = 0$), $M_\uparrow = -M_\downarrow$, so we write $|M_\uparrow| = |M_\downarrow| = M$. We get two equations that are exactly the same:

$$M = g_{\text{eff}}\mu_B J B_J\left(\frac{g_{\text{eff}}\mu_B|\lambda|M}{k_B T}J\right) \tag{5.236}$$

Since this expression is so similar to the expression for ferromagnetism, we simply carry over the results from ferromagnetism and replace λ with $|\lambda|$. We first address the transition temperature:[5]

$$\text{Néel temperature: } T_N = \frac{g_{\text{eff}}^2\mu_B^2|\lambda|J(J+1)}{3k_B} \tag{5.237}$$

We next treat the susceptibility for $T > T_N$, (put back the external field B):

$$M_\uparrow = g_{\text{eff}}\mu_B J B_J\left(\frac{g_{\text{eff}}\mu_B(B - |\lambda|M_\downarrow)}{k_B T}J\right)$$

$$\text{I write } \frac{g_{\text{eff}}\mu_B}{k_B} = \frac{3T_N}{g_{\text{eff}}\mu_B|\lambda|J(J+1)} \quad \text{and expand } B_J(xJ) \approx \frac{J+1}{3J}xJ \tag{5.238}$$

---

[5] Actually, the Néel temperature fits the experimental values badly.

$$M_\uparrow \approx g_{\text{eff}}\mu_B J\left(\frac{J+1}{3J}\frac{3T_N}{g_{\text{eff}}\mu_B|\lambda|J(J+1)}\frac{B-|\lambda|M_\downarrow}{T}J\right) \tag{5.239}$$

$$M_\uparrow = \frac{T_N}{|\lambda|}\frac{B-|\lambda|M_\downarrow}{T}. \tag{5.240}$$

Similarly, the other equation becomes

$$M_\downarrow \approx \frac{T_N}{|\lambda|}\frac{B-|\lambda|M_\uparrow}{T}. \tag{5.241}$$

Solving these two equations gives,

$$M_\uparrow = M_\downarrow = \frac{BT_N}{|\lambda|}\frac{1}{T+T_N} \tag{5.242}$$

$$\implies \chi = \frac{\mu_0 M}{B} = \frac{\mu_0(M_\uparrow+M_\downarrow)}{B} = \frac{2\mu_0 T_N}{|\lambda|}\frac{1}{T+T_N}. \tag{5.243}$$

A more realistic calculation would include the mean fields of the NNNs, but this will not be carried out here.

5.5 Exercises

5.5.1 Question: a numerical comparison of diamagnetism and paramagnetism

(Solution in section 6.5.1.) Estimate the diamagnetic orbital susceptibility of a gas of hydrogen atoms (with number density 1×10^{20} m^{-3}) in the ground state ($n = 1$), and compare this with the paramagnetic spin susceptibility at 100 K.

5.5.2 Question: the exchange interaction is not a magnetic dipole–dipole interaction

(Solution in section 6.5.2.) Estimate the ratio of the exchange and dipolar couplings of two adjacent Fe atoms in metallic Fe. (The exchange constant in Fe can be crudely estimated by setting it equal to $k_B T_C$, where T_C is the Curie temperature. For Fe, $T_C = 1043K$.) The dipolar energy is the usual expression from electromagnetism:

$$E = \frac{\mu_0}{4\pi r^3}\left[\vec{\mu}_1 \cdot \vec{\mu}_2 - \frac{3}{r^2}(\vec{\mu}_1 \cdot \vec{r})(\vec{\mu}_2 \cdot \vec{r})\right]. \tag{5.244}$$

5.5.3 Question: the corrected derivation of ferromagnetic spin waves

(Solution in section 6.5.3.) As a result of imposing a periodic boundary condition on the chain $\uparrow_1\cdots\uparrow_N$, we need to add one term to $H_{\text{Heisenberg}}$. Show that this modification gives a more satisfying derivation of ferromagnetic spin wave dispersion, i.e. we get

$$H_{\text{Heisenberg}}|0\rangle = -\frac{\mathscr{J}N}{4}, \quad H_{\text{Heisenberg}}|q\rangle = \left[-\frac{\mathscr{J}N}{4} + \mathscr{J}\left(1 - \cos(qa)\right)\right]|q\rangle. \quad (5.245)$$

5.5.4 Question: the ground state of an antiferromagnet is a difficult problem

(Solution in section 6.5.4.) Consider the system of four spin-$\frac{1}{2}$ particles with the antiferromagnetic arrangement $|0\rangle = |\uparrow_1 \downarrow_2 \uparrow_3 \downarrow_4\rangle$. This is commonly called the Néel state. Its theoretical model is the antiferromagnetic Heisenberg Hamiltonian $H_{\text{Heisenberg}} = |\mathscr{J}|\sum_{i=1}^{4} \vec{S}_i \cdot \vec{S}_{i+1}$. Use periodic boundary conditions where $\vec{S}_5 = \vec{S}_1$.

1. Show that $|0\rangle$ is not an eigenstate of $H_{\text{Heisenberg}}$. Then show that the expectation value $\langle 0|H_{\text{Heisenberg}}|0\rangle = -|\mathscr{J}|$.
2. The (unnormalised) ground state is

$$|0'\rangle = (|\uparrow_1 \downarrow_2\rangle - |\downarrow_1 \uparrow_2\rangle)(|\uparrow_3 \downarrow_4\rangle - |\downarrow_3 \uparrow_4\rangle) - (|\uparrow_1 \downarrow_4\rangle - |\downarrow_1 \uparrow_4\rangle)(|\uparrow_2 \downarrow_3\rangle - |\downarrow_2 \uparrow_3\rangle)$$
$$|0'\rangle = 2|\uparrow_1 \downarrow_2 \uparrow_3 \downarrow_4\rangle - |\uparrow_1 \downarrow_2 \downarrow_3 \uparrow_4\rangle - |\downarrow_1 \uparrow_2 \uparrow_3 \downarrow_4\rangle$$
$$+ 2|\downarrow_1 \uparrow_2 \downarrow_3 \uparrow_4\rangle - |\uparrow_1 \uparrow_2 \downarrow_3 \downarrow_4\rangle - |\downarrow_1 \downarrow_2 \uparrow_3 \uparrow_4\rangle$$

Check that this is indeed an eigenstate with the eigenvalue $-2|\mathscr{J}|$, i.e. $H_{\text{Heisenberg}}|0'\rangle = -2|\mathscr{J}||0'\rangle$. Do not tackle the 16-dimensional problem directly (see figure 5.16).

5.5.5 Question: the Weiss mean field theory for ferrimagnetism

(Solution in section 6.5.5.) Consider the following model of ferrimagnetism (figure 5.16):

Here, we include next nearest neighbour (nnn) interactions. Let λ denote the mean field constant associated with the exchange constant $\mathscr{J}_{\uparrow\downarrow} = \mathscr{J}_{\downarrow\uparrow}$ and $\lambda_\uparrow(\lambda_\downarrow)$ denote the mean field constant associated with the exchange constant $\mathscr{J}_{\uparrow\uparrow}(\mathscr{J}_{\downarrow\downarrow})$.

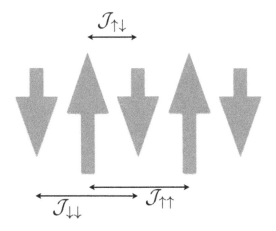

Figure 5.16. The model of ferrimagnetism for question 5.5.5.

1. Write down the two effective fields including the external field B.
2. Write down the Hamiltonian and the magnetisation expressions $M_\uparrow$, $M_\downarrow$. Note that the Landé g-factor and angular momentum J of the two sublattices may not be the same.
3. For $T < T_C$, and no external field,
   (a) Solve for the critical transition temperature T_C by requiring a non-trivial solution for magnetisations $M_\uparrow$, $M_\downarrow$.
   (b) Give a reason why the positive term is chosen to be T_C.
4. For $T > T_C$ and a small external field,
   (a) Solve for the susceptibility $\chi = \frac{\mu_0 M}{B} = \frac{\mu_0}{B}(M_\uparrow + M_\downarrow)$.
5. Treat some special cases: (in each case write down T_C and χ)
   (a) Ferrimagnetism with NN mean field.
   (b) Antiferromagnetism with NNN mean field.
   (c) Antiferromagnetism with NN mean field.
   (d) Ferromagnetism with NN mean field.

References

[1] Kittel C 2005 *Introduction to Solid State Physics* 8th edn (New York: Wiley) https://www.-wiley.com/en-ie/Kittel's+Introduction+to+Solid+State+Physics,+8th+Edition,+Global+Edition-p-9781119454168
[2] Getzlaff M 2007 *Fundamentals of Magnetism* (Berlin: Springer)

Chapter 6

Solutions to the exercises

6.1 Solutions for exercise 1.4

6.1.1 Solution to question 1.4.1

See figure 6.1.

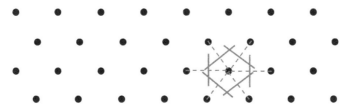

Figure 6.1. Answer for question 1.4.1.

6.1.2 Solution to question 1.4.2

See figure 6.2.

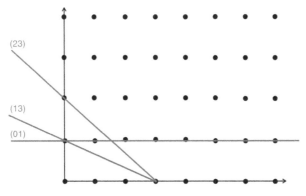

Figure 6.2. Answers for question 1.4.1. For plane (13), it came from the intercepts $3\vec{a}_1$ and $1\vec{a}_2$. For plane (01), it came from the intercepts $\infty\vec{a}_1$ and $1\vec{a}_2$. For plane (23), it came from the intercepts $3\vec{a}_1$ and $2\vec{a}_2$.

6.1.3 Solution to question 1.4.3

See figure 6.3.

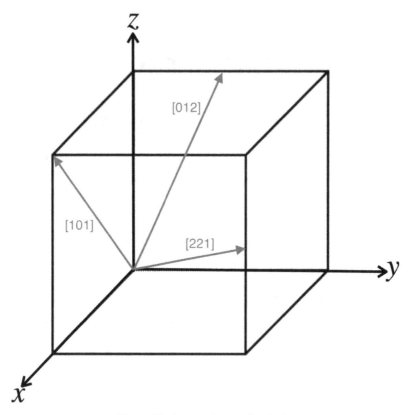

Figure 6.3. Answers for question 1.4.2.

6.1.4 Solution to question 1.4.4

1. • With $\vec{a}_1 = a\hat{x}$ and $\vec{a}_2 = b\hat{y}$, we take $\vec{a}_3 = \hat{z}$.

   • Calculate $|\vec{a}_1 \cdot \vec{a}_2 \times \vec{a}_3| = |a\hat{x} \cdot (b\hat{y} \times \hat{z})| = |a\hat{x} \cdot b\hat{x}| = ab$

   • Then $\vec{b}_1 = 2\pi\frac{\vec{a}_2 \times \vec{a}_3}{ab} = 2\pi\frac{b\hat{y} \times \hat{z}}{ab} = \frac{2\pi}{a}\hat{x}$

   • And $\vec{b}_2 = 2\pi\frac{\vec{a}_3 \times \vec{a}_1}{ab} = 2\pi\frac{\hat{z} \times a\hat{x}}{ab} = \frac{2\pi}{b}\hat{y}$

   • We ignore the calculation of $\vec{b}_3$ as it does not exist (figure 6.4).

2. • Use $d = \frac{2\pi}{|\vec{G}|} = \frac{1}{\sqrt{\frac{4}{a^2} + \frac{9}{b^2}}}$

   • where $\vec{G} = h\vec{b}_1 + k\vec{b}_2 = 2\vec{b}_1 + 3\vec{b}_2 = 2\frac{2\pi}{a}\hat{x} + 3\frac{2\pi}{b}\hat{y}$

6.1.5 Solution to question 1.4.5

See figure 6.4.

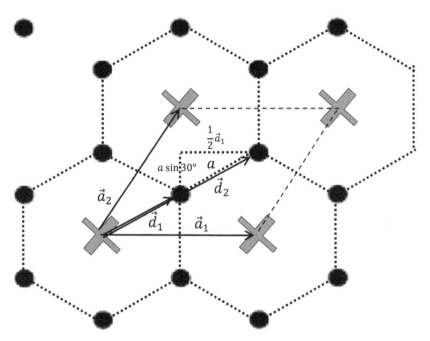

Figure 6.4. The geometry for solution 6.1.5.

1.
- First we need to write down $\vec{a}_1$ and $\vec{a}_2$.
- So note that $\vec{a}_1$ joins the centres of two adjacent hexagons: $\frac{1}{2}\vec{a}_1 = a\cos 30°\hat{x} \implies \vec{a}_1 = a\sqrt{3}\,\hat{x}$.
- For $\vec{a}_2$, we can see that its horizontal component is $\frac{1}{2}\vec{a}_1$ and its vertical component is $\frac{1}{2}a + a\sin 30° + \frac{1}{2}a = \frac{3}{2}a$. So $\vec{a}_2 = \frac{\sqrt{3}}{2}a\hat{x} + \frac{3}{2}a\hat{y}$.
- Therefore, $\vec{d}_1 = \frac{1}{2}\vec{a}_1 + \frac{a}{2}\hat{y} = a\frac{\sqrt{3}}{2}\hat{x} + \frac{a}{2}\hat{y}$.
- And $\vec{d}_2 = \vec{a}_1 + \left(\frac{a}{2} + a\sin 30°\right)\hat{y} = a\sqrt{3}\,\hat{x} + a\hat{y}$

2.
- Let $\vec{a}_3 = \hat{z}$
- Calculate $|\vec{a}_1 \cdot \vec{a}_2 \times \vec{a}_3| = \left| a\sqrt{3}\,\hat{x} \cdot \left(\left(\frac{\sqrt{3}}{2}a\hat{x} + \frac{3}{2}a\hat{y}\right) \times \hat{z}\right)\right| = \frac{3\sqrt{3}}{2}a^2$
- Then $\vec{b}_1 = 2\pi\frac{\vec{a}_2 \times \vec{a}_3}{\frac{3\sqrt{3}}{2}a^2} = 2\pi\frac{\left(-\frac{\sqrt{3}}{2}a\hat{y} + \frac{3}{2}a\hat{x}\right)}{\frac{3\sqrt{3}}{2}a^2} = \frac{2\pi}{a}\left(\frac{1}{\sqrt{3}}\hat{x} - \frac{1}{3}\hat{y}\right)$
- And then $\vec{b}_2 = 2\pi\frac{\vec{a}_3 \times \vec{a}_1}{\frac{3\sqrt{3}}{2}a^2} = 2\pi\frac{\hat{z} \times a\sqrt{3}\,\hat{x}}{\frac{3\sqrt{3}}{2}a^2} = \frac{2\pi}{a}\frac{2}{3}\hat{y}$

6.1.6 Solution to question 1.4.6

- Recall that packing fraction $= \dfrac{\text{volume occupied by spheres}}{\text{volume of hexagonal}}$
- Refering to figure 6.5,

$$\text{volume of hexagonal} = \text{area of hexagonal} \times c \tag{6.1}$$

$$\text{volume of hexagonal} = 6 \times \text{area of triangle} \times c \tag{6.2}$$

$$\text{volume of hexagonal} = 6 \times \frac{1}{2}ah \times c \tag{6.3}$$

$$\text{volume of hexagonal} = 6 \times \frac{1}{2}a\frac{a\sqrt{3}}{2} \times c \tag{6.4}$$

$$\text{volume of hexagonal} = \frac{3\sqrt{3}}{2}a^2c \tag{6.5}$$

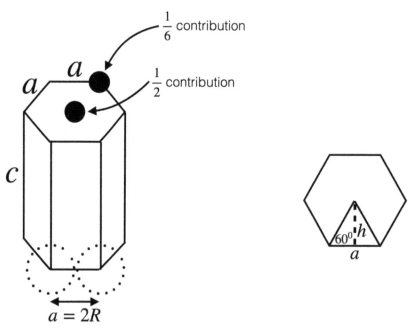

Figure 6,5. The geometry for solution 6.1.6. From the hexagon, we can see that $\tan 30^0 = \frac{\frac{1}{2}a}{h} \implies h = \frac{a}{2\tan 30^0} = \frac{a\sqrt{3}}{2}$.

$$\text{volume occupied by spheres} = \text{number of spheres} \times \frac{4}{3}\pi R^3$$

$$\begin{aligned} &| \text{ corner point contributes } \frac{1}{6} \\ &| \text{ and face point contributes } \frac{1}{2} \\ &| \text{ there are three points inside} \end{aligned} \qquad (6.6)$$

$$\text{volume occupied by spheres} = \left(2 \times 6 \times \frac{1}{6} + 2 \times \frac{1}{2} + 3\right) \times \frac{4}{3}\pi R^3 \qquad (6.7)$$

$$\text{volume occupied by spheres} = 6 \times \frac{4}{3}\pi R^3 \qquad (6.8)$$

$$\underbrace{\text{packing fraction}}_{\frac{\pi\sqrt{2}}{6}} = \frac{6 \times \dfrac{4}{3}\pi R^3}{\dfrac{3\sqrt{3}}{2}a^2 c} \qquad (6.9)$$

$$| \text{ note that } a = 2R$$

$$\frac{\pi\sqrt{2}}{6} = \frac{6 \times \dfrac{4}{3}\pi\left(\dfrac{a}{2}\right)^3}{\dfrac{3\sqrt{3}}{2}a^2 c} \qquad (6.10)$$

$$\frac{c}{a} = \sqrt{\frac{8}{3}} \qquad (6.11)$$

6.1.7 Solution to question 1.4.7

See figure 6.6 for the construction of the Ewald sphere.

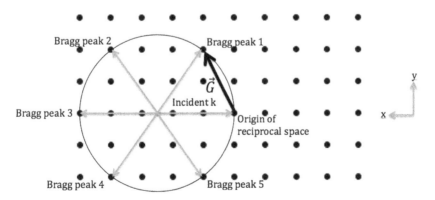

Figure 6.6. The Ewald sphere for solution 6.1.7. The incident k has a magnitude of '2.5 dot spacings'.

- Incident $k = \frac{2\pi}{\lambda} = \frac{2\pi}{\frac{2a}{5}} = \frac{5\pi}{a} = $ '2.5 dot spacings'
- The five Bragg peaks are at these (h, k) coordinates in reciprocal space:
    - Peak 1: (1, 2)
    - Peak 2: (4, 2)
    - Peak 3: (5, 0)
    - Peak 4: (4, −2)
    - Peak 5: (1, −2)
- The (h, k) are five $\vec{G}$ vectors (in units of $\frac{2\pi}{a}$). The planes (in real space) creating these five peaks are normal to these five $\vec{G}$ vectors.

    - Peak 1: (1, 2) $\Longrightarrow$ Plane 1: (12)
    - Peak 2: (4, 2) $\Longrightarrow$ Plane 2 (in the simplest ratio): (21)
    - Peak 3: (5, 0) $\Longrightarrow$ Plane 3 (in the simplest ratio): (10)
    - Peak 4: (4, −2) $\Longrightarrow$ Plane 4 (in the simplest ratio): (2$\bar{1}$)
    - Peak 5: (1, −2) $\Longrightarrow$ Plane 5 (in the simplest ratio): (1$\bar{2}$)
- The Bragg angle 2θ is measured from the incident k. We use the cosine rule to calculate:

$$|\vec{G}|^2 = k^2 + k^2 - 2k^2 \cos 2\theta \Longrightarrow \cos 2\theta = 1 - \frac{|\vec{G}|^2}{2k^2} \qquad (6.12)$$

    - Peak 1: (1, 2) $\Longrightarrow$ Plane 1: (12) $\Longrightarrow$ Bragg angle $2\theta = 53.1^o$.
    - Peak 2: (4, 2) $\Longrightarrow$ Plane 2 (in the simplest ratio): (21) $\Longrightarrow$ Bragg angle $2\theta = 126.9^o$.
    - Peak 3: (5, 0) $\Longrightarrow$ Plane 3 (in the simplest ratio): (10) $\Longrightarrow$ Bragg angle $2\theta = 180^o$.
    - Peak 4: (4, −2) $\Longrightarrow$ Plane 4 (in the simplest ratio): (2$\bar{1}$) $\Longrightarrow$ Bragg angle $2\theta = 126.9^o$.
    - Peak 5: (1, −2) $\Longrightarrow$ Plane 5 (in the simplest ratio): (1$\bar{2}$) $\Longrightarrow$ Bragg angle $2\theta = 53.1^o$.

6.1.8 Solution to question 1.4.8

The structure factor for (real space) BCC is $S_{\vec{G}} = A(1 + (-1)^{h+k+l})$, so

$$S_{\vec{G}} = \begin{cases} 2A \text{ if } h + k + l = \text{even} \\ 0 \text{ if } h + k + l = \text{odd} \end{cases} . \qquad (6.13)$$

We will remove the points in SC reciprocal space where $h + k + l = $ odd. We look at two planes for simplicity. After the points are removed and the planes are stacked together, we get the FCC structure. The number zero counts as even (figure 6.7) (figure 6.8).

6.1.9 Solution to question 1.4.9

The structure factor for (real space) FCC is calculated by treating the real space as SC with a four-atom basis whose displacement vectors are

$(\bar{2}20)$ $(\bar{1}20)$ (020) (120) (220)

$(\bar{2}10)$ $(\bar{1}10)$ (010) (110) (210)

$(\bar{2}00)$ $(\bar{1}00)$ (000) (100) (200)

plane of $l = 0$

$(\bar{2}\bar{1}0)$ $(\bar{1}\bar{1}0)$ $(0\bar{1}0)$ $(1\bar{1}0)$ $(2\bar{1}0)$

$(\bar{2}\bar{2}0)$ $(\bar{1}\bar{2}0)$ $(0\bar{2}0)$ $(1\bar{2}0)$ $(2\bar{2}0)$

Figure 6.7. The $l = 0$ plane of SC reciprocal space with points $h + k + l =$ odd removed.

$(\bar{2}21)$ $(\bar{1}21)$ (021) (121) (221)

$(\bar{2}11)$ $(\bar{1}11)$ (011) (111) (211)

plane of $l = 1$

$(\bar{2}01)$ $(\bar{1}01)$ (001) (101) (201)

$(\bar{2}\bar{1}1)$ $(\bar{1}\bar{1}1)$ $(0\bar{1}1)$ $(1\bar{1}1)$ $(2\bar{1}1)$

$(\bar{2}\bar{2}1)$ $(\bar{1}\bar{2}1)$ $(0\bar{2}1)$ $(1\bar{2}1)$ $(2\bar{2}1)$

Figure 6.8. The $l = 1$ plane of SC reciprocal space with points $h + k + l =$ odd removed.

$$\vec{d}_1 = 0\hat{x} + 0\hat{y} + 0\hat{z} \qquad (6.14)$$

$$\vec{d}_2 = \frac{a}{2}\hat{x} + \frac{a}{2}\hat{y} + 0\hat{z} \qquad (6.15)$$

$$\vec{d}_3 = \frac{a}{2}\hat{x} + 0\hat{y} + \frac{a}{2}\hat{z} \qquad (6.16)$$

$$\vec{d}_4 = 0\hat{x} + \frac{a}{2}\hat{y} + \frac{a}{2}\hat{z}. \tag{6.17}$$

The SC reciprocal space translation vector is

$$\vec{G} = h\vec{b}_1 + k\vec{b}_2 + l\vec{b}_3 = h\frac{2\pi}{a}\hat{x} + k\frac{2\pi}{a}\hat{y} + l\frac{2\pi}{a}\hat{z}. \tag{6.18}$$

The geometrical structure factor is then

$$S_{\vec{G}} = A\left(e^{i\vec{G}\cdot\vec{d}_1} + e^{i\vec{G}\cdot\vec{d}_2} + e^{i\vec{G}\cdot\vec{d}_3} + e^{i\vec{G}\cdot\vec{d}_4}\right) \tag{6.19}$$

$$S_{\vec{G}} = A\left(e^{i0} + e^{\frac{2\pi}{a}\left(h\frac{a}{2}+k\frac{a}{2}\right)} + e^{\frac{2\pi}{a}\left(h\frac{a}{2}+l\frac{a}{2}\right)} + e^{\frac{2\pi}{a}\left(k\frac{a}{2}+l\frac{a}{2}\right)}\right) \tag{6.20}$$

$$S_{\vec{G}} = A(1 + e^{i\pi(h+k)} + e^{i\pi(h+l)} + e^{i\pi(k+l)}) \tag{6.21}$$

$$S_{\vec{G}} = A(1 + (-1)^{h+k} + (-1)^{h+l} + (-1)^{k+l}) \tag{6.22}$$

$$S_{\vec{G}} = \begin{cases} 4A \text{ if } h, k, l \text{ are all even or all odd} \\ 0 \text{ if } h, k, l \text{ are partially even and partially odd} \end{cases}. \tag{6.23}$$

So, using a similar approach to that used for the previous question, we remove points from the $l = 0$ and $l = 1$ planes and stack them together to see the BCC structure of the reciprocal space. The number zero counts as even (figure 6.9) (figure 6.10).

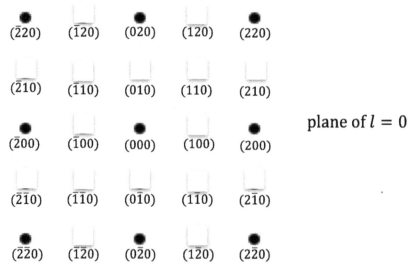

<table>
<tr><td>●</td><td>⊔</td><td>●</td><td>⊔</td><td>●</td></tr>
<tr><td>($\bar{2}$20)</td><td>($\bar{1}$20)</td><td>(020)</td><td>(120)</td><td>(220)</td></tr>
</table>

($\bar{2}$20)	($\bar{1}$20)	(020)	(120)	(220)
($\bar{2}$10)	($\bar{1}$10)	(010)	(110)	(210)
($\bar{2}$00)	($\bar{1}$00)	(000)	(100)	(200)
($\bar{2}\bar{1}$0)	($\bar{1}\bar{1}$0)	(0$\bar{1}$0)	(1$\bar{1}$0)	(2$\bar{1}$0)
($\bar{2}\bar{2}$0)	($\bar{1}\bar{2}$0)	(0$\bar{2}$0)	(1$\bar{2}$0)	(2$\bar{2}$0)

plane of $l = 0$

Figure 6.9. The $l = 0$ plane of SC reciprocal space with points h, k, l = partially even and partially odd removed.

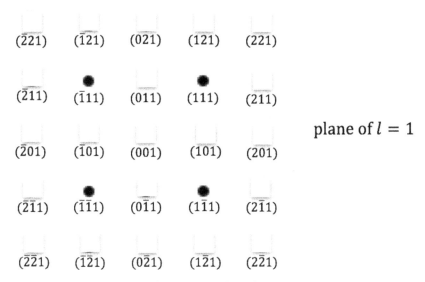

$(\bar{2}21)$ $(\bar{1}21)$ (021) (121) (221)

$(\bar{2}11)$ $(\bar{1}11)$ (011) (111) (211)

$(\bar{2}01)$ $(\bar{1}01)$ (001) (101) (201) **plane of $l = 1$**

$(\bar{2}\bar{1}1)$ $(\bar{1}\bar{1}1)$ $(0\bar{1}1)$ $(1\bar{1}1)$ $(2\bar{1}1)$

$(\bar{2}\bar{2}1)$ $(\bar{1}\bar{2}1)$ $(0\bar{2}1)$ $(1\bar{2}1)$ $(2\bar{2}1)$

Figure 6.10. The $l = 1$ plane of SC reciprocal space with points h, k, l = partially even and partially odd removed.

6.1.10 Solution to question 1.4.10

We need to restrict the general Fourier series $f(\vec{r}) = \sum_{\vec{k}} f(\vec{k}) e^{i\vec{k}\cdot\vec{r}}$ using the condition $f(\vec{r}) = f(\vec{r} + \vec{R})$.

$$f(\vec{r}) = f(\vec{r} + \vec{R}) \tag{6.24}$$
| insert the series into the above condition

$$\sum_{\vec{k}} f(\vec{k}) e^{i\vec{k}\cdot\vec{r}} = \sum_{\vec{k}} f(\vec{k}) e^{i\vec{k}\cdot(\vec{r}+\vec{R})}$$

| use the hint to extract the coefficients for comparison

| so we multiply by $e^{-i\vec{k}'\cdot\vec{r}}$ $\tag{6.25}$

| and integrate over the whole space: $\dfrac{1}{V}\displaystyle\int d^3\vec{r}$

$$\frac{1}{V}\int d^3\vec{r} \sum_{\vec{k}} f(\vec{k}) e^{i(\vec{k}-\vec{k}')\cdot\vec{r}} = \frac{1}{V}\int d^3\vec{r} \sum_{\vec{k}} f(\vec{k}) e^{i(\vec{k}-\vec{k}')\cdot\vec{r}} e^{i\vec{k}\cdot\vec{R}} \tag{6.26}$$

$$\sum_{\vec{k}} f(\vec{k}) \delta_{\vec{k}\vec{k}'} = \sum_{\vec{k}} f(\vec{k}) \delta_{\vec{k}\vec{k}'} e^{i\vec{k}\cdot\vec{R}} \tag{6.27}$$

$$f(\vec{k}') = f(\vec{k}') e^{i\vec{k}'\cdot\vec{R}} \tag{6.28}$$

$$\therefore e^{i\vec{k}'\cdot\vec{R}} = 1 \tag{6.29}$$

$$\therefore \vec{k}' \cdot \vec{R} = 2\pi(\text{integer}) \tag{6.30}$$

$$\therefore \text{allowed } \vec{k} = \vec{G} \tag{6.31}$$

$$\Longrightarrow f(\vec{r}) = \sum_{\vec{G}} f(\vec{G})e^{i\vec{G}\cdot\vec{r}} \tag{6.32}$$

6.1.11 Solution to question 1.4.11

Here, we work out the physics of the Lennard-Jones potential to model crystal binding.

1. Equilibrium separation occurs when the potential energy is at its lowest. Hence, we differentiate and set it to zero to get the minimum point.

$$\frac{dU}{dr}\bigg|_{r=r_0} = 0 \tag{6.33}$$

$$2N\varepsilon\left[A_{12}\left(\frac{-12\sigma^{12}}{r^{13}}\right) - A_6\left(\frac{-6\sigma^6}{r^7}\right)\right]_{r=r_0} = 0 \tag{6.34}$$

$$A_{12}(2\sigma^{12}) - A_6\sigma^6 r_0^6 = 0 \tag{6.35}$$

$$r_0 = \left(\frac{2A_{12}}{A_6}\right)^{1/6} \sigma \tag{6.36}$$

2. Substitute back into $U(r)$ to find the minimum potential energy.

$$U(r_0) = 2N\varepsilon\left[A_{12}\left(\frac{\sigma}{r_0}\right)^{12} - A_6\left(\frac{\sigma}{r_0}\right)^6\right] \tag{6.37}$$

$$U(r_0) = 2N\varepsilon\left[\frac{A_6^2}{4A_{12}} - \frac{A_6^2}{2A_{12}}\right] \tag{6.38}$$

$$U(r_0) = -\frac{N\varepsilon A_6^2}{2A_{12}} \tag{6.39}$$

3. See figure 6.11.

6.1.12 Solution to question 1.4.12

For the ionic solid, we can use an exponential repulsion model:

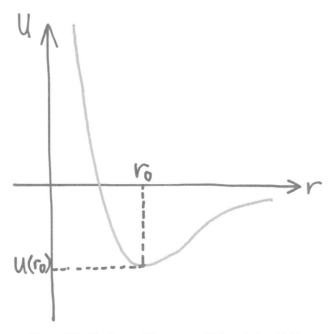

Figure 6.11. The Lennard-Jones potential for solution 6.1.11.

$$U(R) = N\left[z\lambda e^{-R/\rho} - \alpha\frac{q^2}{4\pi\varepsilon_0 R}\right]. \tag{6.40}$$

1. Simply differentiate to get the equilibrium condition

$$\left.\frac{dU}{dR}\right|_{R=R_0} = 0 \tag{6.41}$$

$$\left[Nz\lambda\left(-\frac{1}{\rho}\right)e^{-R/\rho} + N\alpha\frac{q^2}{4\pi\varepsilon_0 R^2}\right]\Bigg|_{R=R_0} = 0 \tag{6.42}$$

$$\frac{\alpha q^2}{4\pi\varepsilon_0 R_0^2} = \frac{z\lambda e^{-R_0/\rho}}{\rho} \tag{6.43}$$

$$R_0^2 e^{-R_0/\rho} = \frac{\rho\alpha q^2}{4\pi\varepsilon_0 z\lambda}. \tag{6.44}$$

2. Substitute back into $U(R)$,

$$U(R_0) = N\left[z\lambda e^{-R_0/\rho} - \alpha\frac{q^2}{4\pi\varepsilon_0 R_0}\right] \tag{6.45}$$

$$U(R_0) = N\left[z\lambda\frac{\rho\alpha q^2}{4\pi\varepsilon_0 R_0^2 z\lambda} - \alpha\frac{q^2}{4\pi\varepsilon_0 R_0}\right] \tag{6.46}$$

$$U(R_0) = -N\frac{\alpha q^2}{4\pi\varepsilon_0 R_0}\left[1 - \frac{\rho}{R_0}\right]. \tag{6.47}$$

6.1.13 Solution to question 1.4.13

For the ionic solid, we can use a power form to model the repulsion:

$$U(R) = N\left[z\lambda\frac{1}{R^n} - \alpha\frac{q^2}{4\pi\varepsilon_0 R}\right]. \tag{6.48}$$

(a) Again, we differentiate to get the equilibrium condition:

$$\left.\frac{dU}{dR}\right|_{R=R_0} = 0 \tag{6.49}$$

$$N\left[z\lambda\left(-\frac{n}{R^{n+1}}\right) + \alpha\frac{q^2}{4\pi\varepsilon_0 R^2}\right]_{R=R_0} = 0 \tag{6.50}$$

$$\frac{\alpha q^2}{4\pi\varepsilon_0 R_0^2} = \frac{z\lambda n}{R_0^{n+1}} \tag{6.51}$$

$$z\lambda = \frac{\alpha q^2}{4\pi\varepsilon_0 n}R_0^{n-1}. \tag{6.52}$$

Substitute back into $U(R)$,

$$U(R_0) = N\left[z\lambda\frac{1}{R_0^n} - \frac{\alpha q^2}{4\pi\varepsilon_0 R_0}\right] \tag{6.53}$$

$$U(R_0) = N\left[\frac{1}{n}\frac{\alpha q^2}{4\pi\varepsilon_0 R_0} - \frac{\alpha q^2}{4\pi\varepsilon_0 R_0}\right] \tag{6.54}$$

$$U(R_0) = -N\frac{\alpha q^2}{4\pi\varepsilon_0 R_0}\left(1 - \frac{1}{n}\right). \tag{6.55}$$

(b)

- Note that $1\ \text{dyne} = 1 \times 10^{-5}\ \text{N}$ so $B = 2.4 \times 10^{11}\ \text{dyne cm}^{-2} = 2.4 \times 10^{10}\ \text{N m}^{-2}$.
- Each cell (see figure 6.12) of volume R^3 contains $\frac{1}{2}$ of Na^+ and $\frac{1}{2}$ of Cl^-.

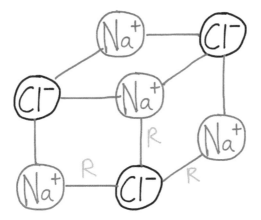

Figure 6.12. A diagram of sodium chloride (NaCl) for solution 6.1.13.

- The volume of the solid is thus $V = 2NR^3$ due to N (the number of) Na^+ and N (the number of) Cl^-.
- At equilibrium, we write $V_0 = 2NR_0^3$.

Calculating the bulk modulus,

$$B_0 = V_0 \left(\frac{d^2U}{dV^2} \right)_{V=V_0}$$

| where $\dfrac{d^2U}{dV^2} = \dfrac{d}{dV}\dfrac{dU}{dV} = \dfrac{dR}{dV}\dfrac{d}{dR}\left(\dfrac{dR}{dV}\dfrac{dU}{dR} \right)$

| resulting in $\dfrac{d^2U}{dV^2} = \dfrac{1}{36NR^2}\left(\dfrac{-\alpha q^2}{\pi\varepsilon_0 R^5} - n(-n-3)z\lambda R^{-n-4} \right)$ $\qquad$ (6.56)

| and use $z\lambda = \dfrac{\alpha q^2}{4\pi\varepsilon_0 n}R_0^{n-1}$

$$B_0 = 2NR_0^3 \frac{1}{36NR_0^2}\left(-\frac{\alpha q^2}{\pi\varepsilon_0 R_0^5} + n(n+3)\frac{\alpha q^2}{4\pi\varepsilon_0 n}R_0^{n-1}R_0^{-n-4} \right) \qquad (6.57)$$

$$B_0 = \left(n - 1 \right)\frac{\alpha q^2}{18(4\pi\varepsilon_0 R_0^4)}. \qquad (6.58)$$

Substitute $B_0 = 2.4 \times 10^{10}$ N m^{-2} and get $n = 7.77 \approx 8$.

6.2 Solutions for exercise 2.3

6.2.1 Solution to question 2.3.1

Simply use the definition of the spring constant:

$$k = \frac{\partial^2 U}{\partial R^2}\bigg|_{R=R_0} \tag{6.59}$$

$$k = \frac{\partial}{\partial R}\left[-\frac{1}{\rho}Nz\lambda e^{-R/\rho} + N\frac{\alpha q^2}{4\pi\varepsilon_0 R^2}\right]_{R=R_0} \tag{6.60}$$

$$k = \frac{1}{\rho^2}Nz\lambda e^{-R_0/\rho} - N\frac{2\alpha q^2}{4\pi\varepsilon_0 R_0^3}$$

$$\bigg| \text{ for equilibrium, use } e^{-R_0/\rho} = \frac{\rho\alpha q^2}{4\pi\varepsilon_0 z\lambda R_0^2} \tag{6.61}$$

$$k = \frac{N\alpha q^2}{4\pi\varepsilon_0 R_0^2}\left(\frac{1}{\rho} - \frac{2}{R_0}\right) \tag{6.62}$$

6.2.2 Solution to question 2.3.2

(a) The dispersion relation is $\omega(q) = \sqrt{\frac{4k}{M}}\left|\sin\frac{qa}{2}\right|$ (figure 6.13).

(b) At the minimum wavelength, adjacent atoms are moving out of phase, hence the minimum wavelength is $2a$. See figure 6.14.

(c) If the wavelength is a,
  - The value of $q = \frac{2\pi}{\lambda} = \frac{2\pi}{a}$.
  - This means all the atoms move together (imagine this using figure 6.15).
  - This is equivalent to $\lambda = \infty$ or $q = 0$.

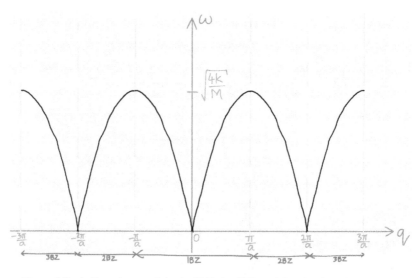

Figure 6.13. A dispersion graph in the 1BZ, the 2BZ, and the 3BZ for solution 6.2.2.

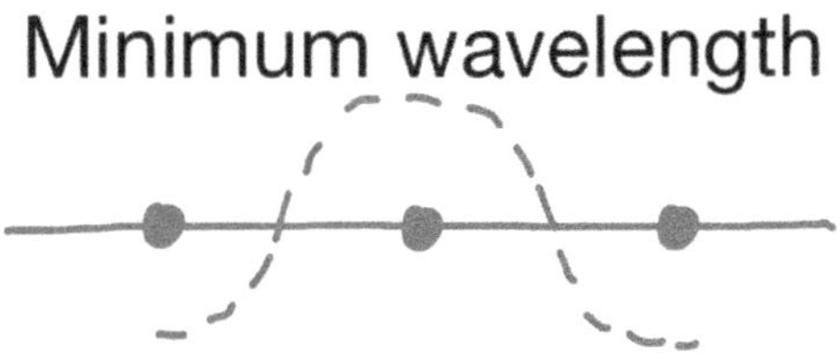

Minimum wavelength

Figure 6.14. The minimum wavelength of a 1D solid with atomic spacing a for solution 6.2.2.

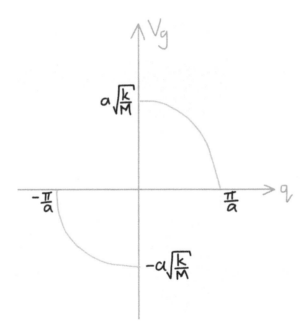

Figure 6.15. A figure illustrating half the wavelength of the minimum wavelength for solution 6.2.2.

Figure 6.16. Group velocity over the 1BZ for solution 6.2.2.

- Indeed, when we translate $q = \frac{2\pi}{a}$ back into the 1BZ using $\vec{G} = h\frac{2\pi}{a}\hat{x}$, we get $q = \frac{2\pi}{a} - \frac{2\pi}{a} = 0$ (figure 6.15).

(d) The group velocity is $v_g = \frac{d\omega}{dq} = \sqrt{\frac{4k}{M}} \frac{a}{2} \cos\left(\frac{qa}{2}\right) = a\sqrt{\frac{k}{M}} \cos\left(\frac{qa}{2}\right)$. The graph for group velocity is shown below. Note that $v_g = 0$ at the BZ boundaries (figure 6.16).

(e) We always follow the rule 'N is the number of atoms (primitive cells actually) and there are N allowed q points' so $N = 7$ means seven allowed q values.

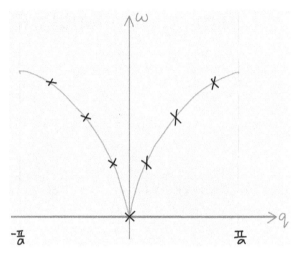

Figure 6.17. There are seven allowed q values, thus there are 'N allowed q points' for solution 6.2.2.

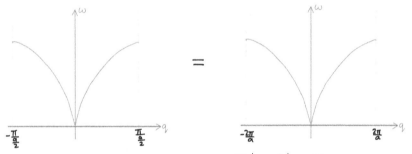

Figure 6.18. The dispersion for atomic spacing $\frac{a}{2}$ is: $\omega(q) = \sqrt{\frac{4k}{M}}\left|\sin\frac{q\frac{a}{2}}{2}\right| = \sqrt{\frac{4k}{M}}\left|\sin\frac{qa}{4}\right|$ for solution 6.2.2.

The allowed q values are: $q = 0, \pm\frac{2\pi}{7a}, \pm\frac{4\pi}{7a}, \pm\frac{6\pi}{7a}$. Figure 6.17 shows the allowed q points on the dispersion graph.

(f)

    i. At $q = 0$, $v_{\text{sound}} = v_g$ so $k = M\left(\frac{v_{\text{sound}}}{a}\right)^2 = 33.8$ N m^{-1}.

    ii. $\omega_{\max} = \sqrt{\frac{4k}{M}} = \frac{2v_{\text{sound}}}{a} = 4.45 \times 10^{13}$ rad s^{-1}.

    iii. In 1D,

- $2q_D = N \times \frac{2\pi}{L_x} \Longrightarrow q_D = \frac{N\pi}{L_x}$
- then $\omega_D = v_{\text{sound}}q_D = v_{\text{sound}}\frac{N\pi}{L_x} = v_{\text{sound}}\frac{\pi}{a}$
- then $\theta_D = \frac{\hbar\omega_D}{k_B} = \frac{\hbar}{k_B}v_{\text{sound}}\frac{\pi}{a} = 535$ K.

(g) We consider

- 1D monoatomic, spacing $\frac{a}{2}$ (figure 6.18):

- 1D diatomic, spacing a but with equal masses: $M_1 = M_2$, so we set $M_1 = M_2 = M$ then (figure 6.19)

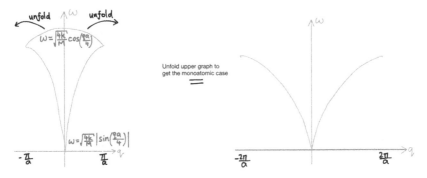

Figure 6.19. For solution 6.2.2: LEFT: diatomic case with equal mass. RIGHT: monoatomic case with spacing $\frac{a}{2}$. The relationship is as follows: unfold the optical branches outwards and we get the monoatomic case.

$$\omega^2 = k\left(\frac{1}{M} + \frac{1}{M}\right) \pm k\sqrt{\left(\frac{1}{M} + \frac{1}{M}\right)^2 - \frac{4}{MM}\sin^2\left(\frac{qa}{2}\right)}$$

$$\omega^2 = \frac{2k}{M} \pm k\sqrt{\frac{4}{M^2} - \frac{4}{M^2}\sin^2\left(\frac{qa}{2}\right)} \tag{6.63}$$

$$\omega^2 = \frac{2k}{M}\left(1 \pm \cos\left(\frac{qa}{2}\right)\right) \tag{6.64}$$

$$\Longrightarrow \begin{cases} \omega^2 = \dfrac{2k}{M}\left(1 + 2\cos^2\left(\dfrac{qa}{4}\right) - 1\right) \\ \omega^2 = \dfrac{2k}{M}\left(1 - \left(1 - 2\sin^2\left(\dfrac{qa}{4}\right)\right)\right) \end{cases} \tag{6.65}$$

$$\Longrightarrow \begin{cases} \omega = \sqrt{\dfrac{4k}{M}}\cos\left(\dfrac{qa}{4}\right) \\ \omega = \sqrt{\dfrac{4k}{M}}\left|\sin\left(\dfrac{qa}{4}\right)\right| \end{cases} \tag{6.66}$$

6.2.3 Solution to question 2.3.3

(a) Photons: $E = \frac{hc}{\lambda} \implies \hbar\omega = \hbar c q \implies \omega = cq$.

(b) Group velocity: $v_g = \frac{d\omega}{dq} = c$ which is the speed of light, as expected.

(c) Wave vector: $q = \frac{2\pi}{\lambda} = 1.57 \times 10^7 \text{ m}^{-1}$.

6.2.4 Solution to question 2.3.4

We propose these quantities for the calculation:

- Debye's model for dispersion: $\omega_{j\vec{q}} = v_j q$
- The two-dimensional density of states (DOS) (from equation (2.68)):

$$D_j(\omega) = \frac{L_x L_y}{2\pi} \frac{q}{v_{gj}} \bigg|_{\omega_{j\vec{q}}=\omega} = \frac{L_x L_y}{2\pi} \frac{\omega_{j\vec{q}}}{v_j^2} \bigg|_{\omega_{j\vec{q}}=\omega} = \frac{L_x L_y}{2\pi} \frac{\omega}{v_j^2} \tag{6.67}$$

- Debye's cutoff frequency ω_D:

$$\pi q_D^2 = N \times \frac{2\pi}{L_x} \frac{2\pi}{L_y} \tag{6.68}$$

$$q_D = \left(N \frac{4\pi}{L_x L_y} \right)^{1/2} \tag{6.69}$$

$$\omega_D = v_j q_D = \left(N \frac{4\pi}{L_x L_y} v_j^2 \right)^{1/2} \tag{6.70}$$

We carry out this calculation after trick one (around equation (2.26)),

$$U = \sum_j \int d\omega D_j(\omega) \frac{\hbar\omega}{e^{\hbar\omega/k_B T} - 1} \tag{6.71}$$

| insert the 2D DOS from above

$$U = \sum_j \int d\omega \frac{L_x L_y}{2\pi} \frac{\omega}{v_j^2} \frac{\hbar\omega}{e^{\hbar\omega/k_B T} - 1} \tag{6.72}$$

| impose the Debye cutoff

$$U = \sum_j \int_0^{\omega_D} d\omega \frac{L_x L_y}{2\pi} \frac{\omega}{v_j^2} \frac{\hbar\omega}{e^{\hbar\omega/k_B T} - 1} \tag{6.73}$$

| assume two branches have the same (constant) group velocity, and label it as v

$$U = \frac{2L_x L_y}{2\pi v^2} \int_0^{\omega_D} d\omega \frac{\hbar\omega^2}{e^{\hbar\omega/k_B T} - 1} \tag{6.74}$$

| let $x = \frac{\hbar\omega}{k_B T}$ and so $x_D = \frac{\hbar_D \omega}{k_B T}$

$$U = \frac{2L_x L_y}{2\pi v^2} \int_0^{x_D} \frac{k_B T}{\hbar} dx \frac{\hbar \left(\frac{k_B T}{\hbar} x \right)^2}{e^x - 1} \tag{6.75}$$

$$= \frac{L_x L_y}{\pi v^2} k_B T \left(\frac{k_B T}{\hbar} \right)^2 \int_0^{x_D} \frac{x^2}{e^x - 1} dx$$

(6.76)

$$| \text{ denote } \Theta_D = \frac{\hbar \omega_D}{k_B} = \frac{\hbar}{k_B} \left(N \frac{4\pi}{L_x L_y} v^2 \right)^{1/2} \implies \frac{L_x L_y}{\pi v^2} = \frac{4 \hbar^2 N}{k_B^2 \Theta_D^2}$$

$$U = 4 k_B T N \left(\frac{T}{\Theta_D} \right)^2 \int_0^{\Theta_D/T} \frac{x^2}{e^x - 1} dx.$$

(6.77)

- For the high-temperature limit, expand e^x:

$$U \approx 4 N k_B T \left(\frac{T}{\Theta_D} \right)^2 \int_0^{\Theta_D} dx \frac{x^2}{1 + x - 1}$$

(6.78)

$$U = 4 N k_B T \left(\frac{T}{\Theta_D} \right)^2 \left[\frac{1}{2} x^2 \right]_0^{\Theta_D/T}$$

(6.79)

$$U = 2 N k_B T$$

(6.80)

$$c_V = 2 N k_B$$

(6.81)

You can call this the Dulong–Petit law in 2D.

- For the low-temperature limit, take $\frac{\Theta_D}{T} \to \infty$:

$$U \approx 4 N k_B T \left(\frac{T}{\Theta_D} \right)^2 \underbrace{\int_0^\infty \frac{x^2}{e^x - 1} dx}_{2.40}$$

(6.82)

$$c_V = 9.6 N k_B \left(\frac{T}{\Theta_D} \right)^2$$

(6.83)

The specific heat has a $\sim T^2$ dependence in 2D.

6.3 Solutions for exercise 3.3

6.3.1 Solution to question 3.3.1

We start with equation (3.73):

$$U(0) = \frac{V}{5\pi^2} \left(\frac{2m}{\hbar^2} \right)^{3/2} \varepsilon_F^{5/2}$$

(6.84)

$$| \text{ use } \varepsilon_F = \frac{\hbar^2}{2m} \left(3\pi^2 \frac{N}{V} \right)^{2/3} \implies \varepsilon_F^{3/2} = \left(\frac{\hbar^2}{2m} \right)^{3/2} 3\pi^2 \frac{N}{V}$$

$$U(0) = \frac{V}{5\pi^2} 3\pi^2 \frac{N}{V} \varepsilon_F$$

(6.85)

$$U(0) = \frac{3}{5} N \varepsilon_F.$$

(6.86)

6.3.2 Solution to question 3.3.2

(a) In 2D, the electrons occupy a Fermi circle. The number of occupied k-points is the 'area of the Fermi circle' divided by 'the area of one k-point'. Each k-point can hold two electrons.

$$N = 2 \times \frac{\pi k_F^2}{\dfrac{2\pi}{L_x} \dfrac{2\pi}{L_y}} \tag{6.87}$$

$$\frac{N}{L_x L_y} = \frac{k_F^2}{2\pi} \tag{6.88}$$

$$\mid \text{let 2D electron density } n = \frac{N}{L_x L_y}$$

$$k_F = \sqrt{2\pi n} \tag{6.89}$$

(b) Use the definition of DOS:

$$D(\varepsilon) = 2 \sum_{\vec{k}} \delta(\varepsilon - \varepsilon_{\vec{k}}) \tag{6.90}$$

$$D(\varepsilon) = 2 \frac{L_x}{2\pi} \frac{L_y}{2\pi} \int \delta\left(\varepsilon - \frac{\hbar^2 k^2}{2m}\right) d^2\vec{k} \tag{6.91}$$

$$\mid \text{use polar coordinates: } d^2\vec{k} = k d\theta dk$$

$$D(\varepsilon) = 2 \frac{L_x}{2\pi} \frac{L_y}{2\pi} \int_0^{2\pi} d\theta \int \delta\left(\varepsilon - \frac{\hbar^2 k^2}{2m}\right) k dk \tag{6.92}$$

$$D(\varepsilon) = \frac{L_x L_y}{\pi} \int \delta\left(\varepsilon - \frac{\hbar^2 k^2}{2m}\right) k dk \tag{6.93}$$

$$\mid \text{let } \varepsilon' = \frac{\hbar^2 k^2}{2m}, \text{ then } d\varepsilon' = \frac{\hbar^2}{m} k dk$$

$$D(\varepsilon) = \frac{m L_x L_y}{\pi \hbar^2} \int \delta(\varepsilon - \varepsilon') d\varepsilon' \tag{6.94}$$

$$D(\varepsilon) = \frac{m L_x L_y}{\pi \hbar^2}. \tag{6.95}$$

(c) Use the definition of N:

$$N = \int_0^{\infty} D(\varepsilon) f(\varepsilon) d\varepsilon \tag{6.96}$$

$$N = \frac{mL_xL_y}{\pi\hbar^2} \int_0^\infty \frac{1}{e^{(\varepsilon-\mu)/k_BT} + 1} d\varepsilon$$

| this integral is exact,

| we evaluate it by letting $x = e^{(\varepsilon-\mu)/k_BT}$ then $dx = \frac{x}{k_BT} d\varepsilon$

| so, $\int_0^\infty \frac{1}{e^{(\varepsilon-\mu)/k_BT} + 1} d\varepsilon = \int_{e^{-\mu/k_BT}}^\infty \frac{k_BT}{x(x + 1)} dx$ (6.97)

| then, using partial fractions, we get $k_BT \int_{e^{-\mu/k_BT}}^\infty \left(\frac{1}{x} - \frac{1}{x + 1}\right) dx$

| evaluating the limits, two infinities cancel each other

| get $k_BT\left(- \ln e^{-\mu/k_BT} + \ln\left(e^{-\mu/k_BT} + 1\right)\right) = k_BT \ln\left(1 + e^{\mu/k_BT}\right)$

$$N = \frac{mL_xL_y}{\pi\hbar^2} k_BT \ln\left(e^{\mu/k_BT} + 1\right)$$ (6.98)

| making μ the subject gives

$$\mu N = k_BT \ln\left(\exp\left(\frac{N\pi\hbar^2}{L_xL_ymk_BT}\right) - 1\right).$$ (6.99)

6.3.3 Solution to question 3.3.3

- In the empty lattice approximation, we have one parabola centred at each k-point. The equation of the parabola is $E = \frac{\hbar^2(\vec{q} - \vec{G}')^2}{2m}$, where $\vec{q}$ is from the 1BZ and $\vec{G}'$ determines the translation to the parabola centred at whichever k-point.
- Every parabola overlaps the 1BZ at various energies. Here, we are required to take into account up to $E = 10\frac{h^2}{8ma^2}$.
- This comes from ten parabolas centred at ten k-points (figure 6.20). We tabulate the calculations in a table. Energies are in units of $\frac{h^2}{8ma^2}$.
- Note that the points have coordinates: $\Gamma = (q_x, q_y) = (0, 0)$, $X = (q_x, q_y) = (\frac{\pi}{a}, 0)$, and $M = (q_x, q_y) = (\frac{\pi}{a}, \frac{\pi}{a})$.

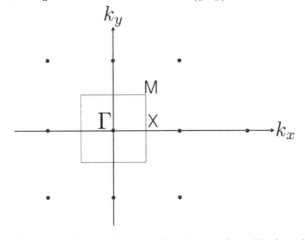

Figure 6.20. For solution 6.3.3: a 2D square BZ with ten reciprocal lattice points shown.

Parabola centred at:	Use this $\vec{G}'$:	Corresponding energy expression:	E_Γ	E_X	E_M
$(k_x, k_y) = (0, 0)$	$\vec{G}' = 0$	$E = \frac{\hbar^2}{2m}\left(q_x^2 + q_y^2\right)$	0	1	2
$(k_x, k_y) = \left(\frac{2\pi}{a}, 0\right)$	$\vec{G}' = \frac{2\pi}{a}\hat{x}$	$E = \frac{\hbar^2}{2m}\left(\left(q_x - \frac{2\pi}{a}\right)^2 + q_y^2\right)$	4	1	2
$(k_x, k_y) = \left(0, \frac{2\pi}{a}\right)$	$\vec{G}' = \frac{2\pi}{a}\hat{y}$	$E = \frac{\hbar^2}{2m}\left(q_x^2 + \left(q_y - \frac{2\pi}{a}\right)^2\right)$	4	5	2
$(k_x, k_y) = \left(0, -\frac{2\pi}{a}\right)$	$\vec{G}' = -\frac{2\pi}{a}\hat{y}$	$E = \frac{\hbar^2}{2m}\left(q_x^2 + \left(q_y + \frac{2\pi}{a}\right)^2\right)$	4	5	10
$(k_x, k_y) = \left(-\frac{2\pi}{a}, 0\right)$	$\vec{G}' = \frac{2\pi}{a}\hat{x}$	$E = \frac{\hbar^2}{2m}\left(\left(q_x + \frac{2\pi}{a}\right)^2 + q_y^2\right)$	4	9	10
$(k_x, k_y) = \left(\frac{2\pi}{a}, \frac{2\pi}{a}\right)$	$\vec{G}' = \frac{2\pi}{a}\hat{x} + \frac{2\pi}{a}\hat{y}$	$E = \frac{\hbar^2}{2m}\left(\left(q_x - \frac{2\pi}{a}\right)^2 + \left(q_y - \frac{2\pi}{a}\right)^2\right)$	8	5	2
$(k_x, k_y) = \left(-\frac{2\pi}{a}, \frac{2\pi}{a}\right)$	$\vec{G}' = -\frac{2\pi}{a}\hat{x} + \frac{2\pi}{a}\hat{y}$	$E = \frac{\hbar^2}{2m}\left(\left(q_x + \frac{2\pi}{a}\right)^2 + \left(q_y - \frac{2\pi}{a}\right)^2\right)$	8	13	10
$(k_x, k_y) = \left(\frac{2\pi}{a}, -\frac{2\pi}{a}\right)$	$\vec{G}' = \frac{2\pi}{a}\hat{x} - \frac{2\pi}{a}\hat{y}$	$E = \frac{\hbar^2}{2m}\left(\left(q_x - \frac{2\pi}{a}\right)^2 + \left(q_y + \frac{2\pi}{a}\right)^2\right)$	8	5	10
$(k_x, k_y) = \left(-\frac{2\pi}{a}, -\frac{2\pi}{a}\right)$	$\vec{G}' = -\frac{2\pi}{a}\hat{x} - \frac{2\pi}{a}\hat{y}$	$E = \frac{\hbar^2}{2m}\left(\left(q_x + \frac{2\pi}{a}\right)^2 + \left(q_y + \frac{2\pi}{a}\right)^2\right)$	8	13	18
$(k_x, k_y) = \left(\frac{4\pi}{a}, 0\right)$	$\vec{G}' = \frac{4\pi}{a}\hat{x}$	$E = \frac{\hbar^2}{2m}\left(\left(q_x - \frac{4\pi}{a}\right)^2 + q_y^2\right)$	16	9	10

The dispersion is sketched in figure 6.21.

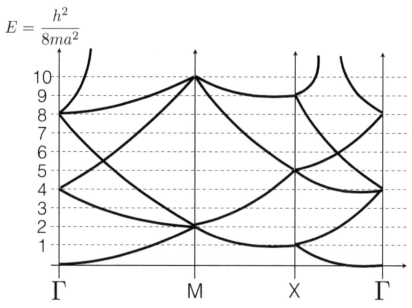

Figure 6.21. For solution 6.3.3: a dispersion graph in the empty lattice approximation up to $E = 10\frac{\hbar^2}{8ma^2}$. Animation available at https://doi.org/10.1088/978-0-7503-5217-8.

6.3.4 Solution to question 3.3.4

(a) You can also see Q3 for reference.

- We have $E = \frac{\hbar^2}{2m}(\vec{q} - \vec{G'})^2$ and for the lowest band $\vec{G'} = 0$.

- So $E = \frac{\hbar^2}{2m}(q_x^2 + q_y^2)$. Then points $X = (q_x = \frac{\pi}{a}, q_y = 0)$ and $M = (q_x = \frac{\pi}{a}, q_y = \frac{\pi}{a})$.

- So $E_X = \frac{\hbar^2}{2m}\left(\left(\frac{\pi}{a}\right)^2 + 0^2\right)$ and $E_M = \frac{\hbar^2}{2m}\left(\left(\frac{\pi}{a}\right)^2 + \left(\frac{\pi}{a}\right)^2\right)$

- Therefore $E_M = 2E_X$.

(b) We simply need to write it into complex Fourier series and identify the Fourier coefficients.

$$V(x, y) = 2V_0\left(\cos\left(\frac{2\pi x}{a}\right) + \cos\left(\frac{2\pi y}{a}\right)\right) \tag{6.100}$$

$$= V_0\left(e^{i\frac{2\pi}{a}x} + e^{-i\frac{2\pi}{a}x} + e^{i\frac{2\pi}{a}y} + e^{-i\frac{2\pi}{a}y}\right) \tag{6.101}$$

The overlap at X is between the parabola centred at $(k_x = 0, k_y = 0)$ and the parabola centred at $(k_x = \frac{2\pi}{a}, k_y = 0)$. Thus, the bandgap at X is $2|V_{2\pi/a}| = 2V_0$.

(c) If we can tune V_0 up so that band two is high enough to be higher than E_M, then there is no band overlap and it becomes an insulator. Note that for this potential, there is only a bandgap when nearest parabolas overlap.

6.3.5 Solution to question 3.3.5

(a) The 1D tight binding dispersion is $\varepsilon_{\vec{k}} = E^{\text{atom}} - E^{\text{onsite}} - 2ta\cos(ka)$.
(b) The band width is the difference between the highest energy in the band and the lowest energy in the band: $(E^{\text{atom}} - E^{\text{onsite}} + 2ta) - (E^{\text{atom}} - E^{\text{onsite}} - 2ta) = 4ta$.
(c) The group velocity is exactly the sine graph.

6.3.6 Solution to question 3.3.6

(a) For the NFE model, near the BZ edge, $(q \approx \frac{\pi}{a})$,

- Consider the first–second band intersection: $\vec{G}_a' = 0$ and $\vec{G}_b' = \frac{2\pi}{a}\hat{x}$.
- We shall look at the (sad-face) parabola in the first band. Thus, we take the $-$ sign.
- Write $q \approx \frac{\pi}{a} + \delta$ where δ is small and negative.

$$E = \frac{1}{2}\left[\frac{\hbar^2}{2m}\left(\frac{\pi}{a} + \delta\right)^2 + \frac{\hbar^2}{2m}\left(\frac{\pi}{a} + \delta - \frac{2\pi}{a}\right)^2 \right]$$
$$- \sqrt{\left(\frac{\frac{\hbar^2}{2m}\left(\frac{\pi}{a} + \delta\right)^2 - \frac{\hbar^2}{2m}\left(\frac{\pi}{a} + \delta - \frac{2\pi}{a}\right)^2}{2} \right)^2 + |V_{2\pi/a}|^2} \qquad (6.102)$$

| so $\dfrac{\hbar^2}{2m}\left(\dfrac{\pi}{a} + \delta\right)^2 + \dfrac{\hbar^2}{2m}\left(\dfrac{\pi}{a} + \delta - \dfrac{2\pi}{a}\right)^2 = \dfrac{2\pi^2}{a^2} + 2\delta^2$

| so $\dfrac{\hbar^2}{2m}\left(\dfrac{\pi}{a} + \delta\right)^2 - \dfrac{\hbar^2}{2m}\left(\dfrac{\pi}{a} + \delta - \dfrac{2\pi}{a}\right)^2 = \dfrac{4\pi}{a}\delta$

$$E = \frac{\hbar^2}{2m}\left(\frac{\pi^2}{a^2} + \delta^2\right) - \sqrt{\left(\frac{\hbar^2}{2m}\frac{2\pi}{a}\delta\right)^2 + |V_{2\pi/a}|^2} \qquad (6.103)$$

$$E = \frac{\hbar^2}{2m}\left(\frac{\pi}{a}\right)^2 + \frac{\hbar^2}{2m}\delta^2 - |V_{2\pi/a}|\sqrt{1 + \frac{4\frac{\hbar^2}{2m}\left(\frac{\pi}{a}\right)^2\frac{\hbar^2}{2m}\delta^2}{|V_{2\pi/a}|^2}} \qquad (6.104)$$

| binomial expand the square root, so $\sqrt{1 + x} \approx 1 + \dfrac{1}{2}x$ where $x \ll 1$

$$E \approx \frac{\hbar^2}{2m}\left(\frac{\pi}{a}\right)^2 + \frac{\hbar^2}{2m}\delta^2 - |V_{2\pi/a}|\left(1 + \frac{1}{2}\frac{4\frac{\hbar^2}{2m}\left(\frac{\pi}{a}\right)^2\frac{\hbar^2}{2m}\delta^2}{|V_{2\pi/a}|^2}\right) \qquad (6.105)$$

$$E = \frac{\hbar^2}{2m}\left(\frac{\pi}{a}\right)^2 - |V_{2\pi/a}| + \left(1 - \frac{2\frac{\hbar^2}{2m}\left(\frac{\pi}{a}\right)^2}{|V_{2\pi/a}|}\right)\frac{\hbar^2}{2m}\delta^2 \qquad (6.106)$$

So this is a parabola in δ. The term $\frac{\hbar^2}{2m}\left(\frac{\pi}{a}\right)^2 - |V_{2\pi/a}|$ is the 'y-intercept' of the parabola. The factor $\left(1 - \frac{2\frac{\hbar^2}{2m}\left(\frac{\pi}{a}\right)^2}{|V_{2\pi/a}|}\right)$ is negative (sad-face parabola) and can be grouped to find the effective mass.

(b) For 1D, we can immediately deduce $E = E^{\text{atom}} - E^{\text{onsite}} - 2ta\cos(ka)$.

- Near the BZ centre: $k \ll 1$, so $\cos(ka) \approx 1 - \frac{1}{2}k^2a^2$; then $E \approx E^{\text{atom}} - E^{\text{onsite}} - 2ta + ta^3k^2$ which is a smiley parabola.
- Near the BZ edge: $k \approx \frac{\pi}{a} + \delta$ where δ is small and negative.

$$E = E^{\text{atom}} - E^{\text{onsite}} - 2ta\cos(\pi + \delta a) \tag{6.107}$$

$$E = E^{\text{atom}} - E^{\text{onsite}} - 2ta(\cos\pi\cos\delta a - \sin\pi\sin\delta a) \tag{6.108}$$

$$E = E^{\text{atom}} - E^{\text{onsite}} + 2ta\cos\delta a \tag{6.109}$$

$$E \approx E^{\text{atom}} - E^{\text{onsite}} + 2ta\left(1 - \frac{1}{2}\delta^2a^2\right) \tag{6.110}$$

$$E = E^{\text{atom}} - E^{\text{onsite}} + 2ta - ta^3\delta^2 \tag{6.111}$$

which is a sad-face parabola.

6.3.7 Solution to question 3.3.7

(a)

- Boundary condition 1: $\psi_I(0) = \psi_{II}(0)$

$$Ae^{iq0} + Be^{-iq0} = Ce^{Q0} + De^{-Q0} \tag{6.112}$$

$$\implies A + B = C + D \tag{6.113}$$

- Boundary condition 2:

$$\left.\frac{d\psi_I}{dx}\right|_{x=0} = \left.\frac{d\psi_{II}}{dx}\right|_{x=0} \tag{6.114}$$

$$\implies Aiq - Biq = CQ - DQ \tag{6.115}$$

- Boundary condition 3:

$$\psi_{II}(b) = e^{ika}\psi_I(-w) \tag{6.116}$$

$$\implies Ce^{Qb} + De^{-Qb} = e^{ika}(Ae^{-iqw} + Be^{iqw}) \tag{6.117}$$

- Boundary condition 4:

$$\left.\frac{d\psi_{II}}{dx}\right|_{x=b} = e^{ika}\left.\frac{d\psi_I}{dx}\right|_{x=-w} \tag{6.118}$$

$$\implies CQe^{Qb} - DQe^{-Qb} = e^{ika}(Aiqe^{-iqw} - Biqe^{iqw}) \tag{6.119}$$

(b) Forming the system of linear equations:

$$\begin{pmatrix} 1 & 1 & -1 & -1 \\ iq & -iq & -Q & Q \\ e^{-iqw+ika} & e^{iqw+ika} & -e^{Qb} & -e^{-Qb} \\ iqe^{-iqw+ika} & -iqe^{iqw+ika} & -Qe^{Qb} & Qe^{-Qb} \end{pmatrix} \begin{pmatrix} A \\ B \\ C \\ D \end{pmatrix} = 0 \qquad (6.120)$$

Now set the determinant of the coefficient matrix to zero, then use Laplace expansion taking the last row.

$$0 = \begin{vmatrix} 1 & 1 & -1 & -1 \\ iq & -iq & -Q & Q \\ e^{-iqw+ika} & e^{iqw+ika} & -e^{Qb} & -e^{-Qb} \\ iqe^{-iqw+ika} & -iqe^{iqw+ika} & -Qe^{Qb} & Qe^{-Qb} \end{vmatrix} \qquad (6.121)$$

$$0 = iqe^{-iqw+ika} \begin{vmatrix} 1 & -1 & -1 \\ -iq & -Q & Q \\ e^{iqw+ika} & -e^{Qb} & -e^{-Qb} \end{vmatrix}$$

$$+ iqe^{iqw+ika} \begin{vmatrix} 1 & -1 & -1 \\ iq & -Q & Q \\ e^{-iqw+ika} & -e^{Qb} & -e^{-Qb} \end{vmatrix}$$

$$- Qe^{Qb} \begin{vmatrix} 1 & 1 & -1 \\ iq & -iq & Q \\ e^{-iqw+ika} & e^{iqw+ika} & -e^{-Qb} \end{vmatrix}$$

$$- Qe^{-Qb} \begin{vmatrix} 1 & 1 & -1 \\ iq & -iq & -Q \\ e^{-iqw+ika} & e^{iqw+ika} & -e^{Qb} \end{vmatrix} \qquad (6.122)$$

We work out these four determinants separately.

$$\text{First determinant} = \begin{vmatrix} 1 & -1 & -1 \\ -iq & -Q & Q \\ e^{iqw+ika} & -e^{Qb} & -e^{-Qb} \end{vmatrix} \qquad (6.123)$$

$$\text{First determinant} = e^{iqw+ika} \begin{vmatrix} -1 & -1 \\ -Q & Q \end{vmatrix} + e^{Qb} \begin{vmatrix} 1 & -1 \\ -iq & Q \end{vmatrix}$$

$$- e^{-Qb} \begin{vmatrix} 1 & -1 \\ -iq & -Q \end{vmatrix} \qquad (6.124)$$

$$\text{First determinant} = e^{iqw+ika}(-2Q) + e^{Qb}(Q - iq)$$
$$- e^{-Qb}(-Q - iq) \qquad (6.125)$$

First determinant $= -2Qe^{iqw+ika} + 2Q\cosh(Qb) - 2iq\sinh(Qb)$ (6.126)

$$\text{Second determinant} = \begin{vmatrix} 1 & -1 & -1 \\ iq & -Q & Q \\ e^{-iqw+ika} & -e^{Qb} & -e^{-Qb} \end{vmatrix}$$ (6.127)

$$\text{Second determinant} = e^{-iqw+ika}\begin{vmatrix} -1 & -1 \\ -Q & Q \end{vmatrix} + e^{Qb}\begin{vmatrix} 1 & -1 \\ iq & Q \end{vmatrix}$$

$$- e^{-Qb}\begin{vmatrix} 1 & -1 \\ iq & -Q \end{vmatrix}$$ (6.128)

Second determinant $= e^{-iqw+ika}(-2Q) + e^{Qb}(Q+iq)$

$- e^{-Qb}(-Q+iq)$ (6.129)

Second determinant $= -2Qe^{-iqw+ika} + 2Q\cosh(Qb)$

$+ 2iq\sinh(Qb)$ (6.130)

$$\text{Third determinant} = \begin{vmatrix} 1 & 1 & -1 \\ iq & -iq & Q \\ e^{-iqw+ika} & e^{iqw+ika} & -e^{-Qb} \end{vmatrix}$$ (6.131)

$$\text{Third determinant} = e^{-iqw+ika}\begin{vmatrix} 1 & -1 \\ -iq & Q \end{vmatrix} - e^{iqw+ika}\begin{vmatrix} 1 & -1 \\ iq & Q \end{vmatrix}$$

$$- e^{-Qb}\begin{vmatrix} 1 & 1 \\ iq & -iq \end{vmatrix}$$ (6.132)

Third determinant $= e^{-iqw+ika}(Q-iq) - e^{iqw+ika}(Q+iq)$

$- e^{-Qb}(-iq-iq)$ (6.133)

Third determinant $= -2iQe^{ika}\sin(qw) - 2iqe^{ika}\cos(qw)$

$+ 2iqe^{-Qb}$ (6.134)

$$\text{Fourth determinant} = \begin{vmatrix} 1 & 1 & -1 \\ iq & -iq & -Q \\ e^{-iqw+ika} & e^{iqw+ika} & -e^{Qb} \end{vmatrix}$$ (6.135)

$$\text{Fourth determinant} = e^{-iqw+ika}\begin{vmatrix} 1 & -1 \\ -iq & -Q \end{vmatrix} - e^{iqw+ika}\begin{vmatrix} 1 & -1 \\ iq & -Q \end{vmatrix}$$

$$- e^{Qb}\begin{vmatrix} 1 & 1 \\ iq & -iq \end{vmatrix}$$ (6.136)

$$\text{Fourth determinant} = e^{-iqw+ika}(-Q - iq) - e^{iqw+ika}(-Q + iq) \\ + 2iqe^{Qb} \tag{6.137}$$

$$\text{Fourth determinant} = 2iQe^{ika}\sin(qw) - 2iqe^{ika}\cos(qw) + 2iqe^{Qb} \tag{6.138}$$

Putting everything back:

$$0 = iqe^{-iqw+ika}[-2Qe^{iqw+ika} + 2Q\cosh(Qb) - 2iq\sinh(Qb)] \\ + iqe^{iqw+ika}[-2Qe^{-iqw+ika} + 2Q\cosh(Qb) + 2iq\sinh(Qb)] \\ - Qe^{Qb}[-2iQe^{ika}\sin(qw) - 2iqe^{ika}\cos(qw) + 2iqe^{-Qb}] \\ - Qe^{-Qb}[2iQe^{ika}\sin(qw) - 2iqe^{ika}\cos(qw) + 2iqe^{Qb}] \tag{6.139}$$

| expand and find that all terms add pairwise
| then multiply e^{-ika} on both sides

$$0 = -4iqQe^{ika} + 4iqQ\cos(qw)\cosh(Qb) - 4iq^2\sinh(Qb)\sin(qw) \\ + 4iQ^2\sinh(Qb)\sin(qw) + 4iqQ\cosh(Qb)\cos(qw) \\ - 4iqQe^{-ika} \tag{6.140}$$

$$0 = -8iqQ\cos(ka) + 81qQ\cosh(Qb)\cos(qw) \\ + 4i(Q^2 - q^2)\sinh(Qb)\sin(qw) \tag{6.141}$$

| divide by $81qQ$ on both sides and move $\cos(ka)$ to the other side

$$\cos(ka)0 = \frac{Q^2 - q^2}{2qQ}\sinh(Qb)\sin(qw) + \cosh(Qb)\cos(qw) \tag{6.142}$$

(c) Recall that $Q^2 = \frac{2m}{\hbar^2}(V_0 - E)$ and $q^2 = \frac{2m}{\hbar^2}E$. In the Dirac delta function barriers limit, $b \to 0$ so $w \to a$ but area $V_0 b$ is constant.

$$\frac{Q^2 - q^2}{2qQ}\underbrace{\sinh(Qb)}_{\approx Qb}\underbrace{\sin(qw)}_{\approx\sin(qa)} + \underbrace{\cosh(Qb)}_{\approx 1}\underbrace{\cos(qw)}_{\approx\cos(qa)} = \cos(ka) \tag{6.143}$$

$$\frac{Q^2 - q^2}{2q}b\sin(qa) + \cos(qa) = \cos(ka)$$

$$\underbrace{V_0 \gg E}_{\text{Dirac potential}} \implies \frac{Q^2 - q^2}{2q} = \frac{mV_0}{\hbar^2 q} - \frac{2mE}{\hbar^2} \approx \frac{mV_0}{\hbar^2 q} \quad | \tag{6.144}$$

$$\frac{mV_0 b}{\hbar^2 q}\sin(qa) + \cos(qa) = \cos(ka) \tag{6.145}$$

$$\frac{mV_0ba}{\hbar^2}\frac{\sin(qa)}{qa} + \cos(qa) = \cos(ka) \qquad (6.146)$$

6.3.8 Solution to question 3.3.8

1. The 3D tight binding dispersion is

$$\varepsilon_{\vec{k}} = E^{\text{atom}} - E^{\text{onsite}} - 2t_x a \cos(k_x a) - 2t_y b \cos(k_y b) - 2t_z c \cos(k_z c) \qquad (6.147)$$

and the definition of (inverse) effective mass is

$$(m^{*-1})_{ij} = \frac{1}{\hbar^2}\frac{\partial^2 \varepsilon_{\vec{k}}}{\partial k_i \partial k_j} \qquad (6.148)$$

$$m^{*-1} = \begin{pmatrix} \dfrac{1}{\hbar^2}\dfrac{\partial^2 \varepsilon_{\vec{k}}}{\partial k_x^2} & \dfrac{1}{\hbar^2}\dfrac{\partial^2 \varepsilon_{\vec{k}}}{\partial k_x \partial k_y} & \dfrac{1}{\hbar^2}\dfrac{\partial^2 \varepsilon_{\vec{k}}}{\partial k_x \partial k_z} \\ \dfrac{1}{\hbar^2}\dfrac{\partial^2 \varepsilon_{\vec{k}}}{\partial k_y \partial k_x} & \dfrac{1}{\hbar^2}\dfrac{\partial^2 \varepsilon_{\vec{k}}}{\partial k_y^2} & \dfrac{1}{\hbar^2}\dfrac{\partial^2 \varepsilon_{\vec{k}}}{\partial k_y \partial k_z} \\ \dfrac{1}{\hbar^2}\dfrac{\partial^2 \varepsilon_{\vec{k}}}{\partial k_z \partial k_x} & \dfrac{1}{\hbar^2}\dfrac{\partial^2 \varepsilon_{\vec{k}}}{\partial k_z \partial k_y} & \dfrac{1}{\hbar^2}\dfrac{\partial^2 \varepsilon_{\vec{k}}}{\partial k_z^2} \end{pmatrix} \qquad (6.149)$$

$$m^{*-1} = \begin{pmatrix} \dfrac{2t_x a^2}{\hbar^2}\cos(k_x a) & 0 & 0 \\ 0 & \dfrac{2t_y b^2}{\hbar^2}\cos(k_y b) & 0 \\ 0 & 0 & \dfrac{2t_z c^2}{\hbar^2}\cos(k_z c) \end{pmatrix} \qquad (6.150)$$

| invert the diagonal matrix

$$m^{*} = \begin{pmatrix} \dfrac{\hbar^2}{2t_x a^2 \cos(k_x a)} & 0 & 0 \\ 0 & \dfrac{\hbar^2}{2t_y b^2 \cos(k_y b)} & 0 \\ 0 & 0 & \dfrac{\hbar^2}{2t_z c^2 \cos(k_z c)} \end{pmatrix} \qquad (6.151)$$

2. In this case,

$$m^{*} = \frac{\hbar^2}{2ta^2 \cos(ka)}\begin{pmatrix} 1 & 0 & 0 \\ 0 & 1 & 0 \\ 0 & 0 & 1 \end{pmatrix} \qquad (6.152)$$

6.4 Solutions for exercise 4.3

6.4.1 Solution to question 4.3.1

We start with the general equation for a triaxial ellipsoid centred at $(k_{x_0}, k_{y_0}, k_{z_0})$:

$$\frac{(k_x - k_{x_0})^2}{d^2} + \frac{(k_y - k_{y_0})^2}{b^2} + \frac{(k_z - k_{z_0})^2}{c^2} = 1. \tag{6.153}$$

In this case, we have $k_{x_0} = 0$, $k_{y_0} = 0$, $k_{z_0} = \frac{3}{4}\frac{2\pi}{a}$, and $d = b$ because of circular cross sections. So now the equation for the spheroid is

$$\frac{k_x^2 + k_y^2}{b^2} + \frac{\left(k_z - \frac{3}{4}\frac{2\pi}{a}\right)^2}{c^2} = 1. \tag{6.154}$$

We compare with a spherical constant energy surface $\varepsilon = \frac{\hbar^2 k^2}{2m^*} \Longrightarrow 1 = \frac{\hbar^2}{2m^*\varepsilon}k^2$ and hence introduce two effective masses, as follows: $\frac{1}{b^2} = \frac{\hbar^2}{2m^*_{(x \text{ or } y)}\varepsilon}$ and $\frac{1}{c^2} = \frac{\hbar^2}{2m^*_{(z)}\varepsilon}$. Hence, the dispersion for spheroid is:

$$\varepsilon = \frac{\hbar^2}{2}\left(\frac{k_x^2 + k_y^2}{m^*_{(x \text{ or } y)}} + \frac{\left(k_z - \frac{3}{4}\frac{2\pi}{a}\right)^2}{m^*_{(z)}}\right). \tag{6.155}$$

6.4.2 Solution to question 4.3.2

From quantum mechanics textbooks, we solve the Schrödinger equation in relative coordinates:

$$E_n = -\frac{\mu}{2\hbar^2}\left(\frac{e^2}{4\pi\varepsilon_0}\right)^2\frac{1}{n^2}$$

$$\left|\begin{array}{l} \text{where } \mu \text{ is the reduced mass: } \frac{1}{\mu} = \frac{1}{m^*_c} + \frac{1}{m^*_v} \\ \text{modify for exciton: } \varepsilon_0 \Longrightarrow \varepsilon_0\varepsilon_r \\ \text{let } n = 1 \text{ for ground state} \end{array}\right. \tag{6.156}$$

$$E_{\text{exciton}} = E_1 = -\frac{m^*_c m^*_v}{m^*_c + m^*_v}\frac{1}{\varepsilon_r^2}\frac{1}{2\hbar^2}\left(\frac{e^2}{4\pi\varepsilon_0}\right)^2$$

$$\left|\text{ insert } \frac{m_e}{m_e} = 1 \text{ and note that } \frac{m_e}{2\hbar^2}\left(\frac{e^2}{4\pi\varepsilon_0}\right)^2 = 13.6\,eV\right. \tag{6.157}$$

$$E_{\text{exciton}} = -\frac{m_c^* m_v^*}{m_e(m_c^* + m_v^*)} \frac{1}{\varepsilon_r^2} \times 13.6 \text{ eV}. \tag{6.158}$$

6.4.3 Solution to question 4.3.3

Just proceed with a very similar calculation:

$$p_v = \frac{1}{V} \int_{-\infty}^{E_v} D(\varepsilon)(1 - f(\varepsilon))d\varepsilon \tag{6.159}$$

$$p_v = \frac{1}{V} \int_{-\infty}^{E_v} D(\varepsilon) \frac{1}{e^{(\mu-\varepsilon)/k_B T} + 1} d\varepsilon$$

| approximate $\mu - E_v \gg k_B T$, so $\dfrac{1}{e^{(\mu-\varepsilon)/k_B T} + 1} \approx e^{-(\mu-\varepsilon)/k_B T}$ \qquad (6.160)

| and put $1 = e^{E_v/k_B T} e^{-E_v/k_B T}$

$$p_v \approx \frac{1}{V} \int_{-\infty}^{E_v} D(\varepsilon) e^{E_v/k_B T} e^{-E_v/k_B T} e^{-(\mu-\varepsilon)/k_B T} d\varepsilon \tag{6.161}$$

$$p_v = \frac{1}{V} e^{(E_v-\mu)/k_B T} \int_{-\infty}^{E_v} D(\varepsilon) e^{(\varepsilon-E_v)/k_B T} d\varepsilon$$

| the DOS is $D(\varepsilon) = \dfrac{V}{2\pi^2} \left(\dfrac{2m_v^*}{\hbar^2}\right)^{3/2} \sqrt{E_v - \varepsilon}$ \qquad (6.162)

$$p_v = \frac{1}{2\pi^2} e^{(E_v-\mu)/k_B T} \left(\frac{2m_v^*}{\hbar^2}\right)^{3/2} \int_{-\infty}^{E_v} \sqrt{E_v - \varepsilon}\, e^{(\varepsilon-E_v)/k_B T} d\varepsilon$$

| let $\varepsilon' = \dfrac{E_v - \varepsilon}{k_B T}$ and $-k_B T d\varepsilon' = d\varepsilon$ \qquad (6.163)

$$p_v = \frac{1}{2\pi^2} e^{(E_v-\mu)/k_B T} \left(\frac{2m_v^*}{\hbar^2}\right)^{3/2} (-(k_B T)^{3/2}) \int_{\infty}^{0} \sqrt{\varepsilon'}\, e^{-\varepsilon'} d\varepsilon'$$

| use integral identity $\displaystyle\int_0^{\infty} \sqrt{x}\, e^{-x} dx = \frac{\sqrt{\pi}}{2}$ \qquad (6.164)

$$p_v = 2 \left(\frac{m_v^* k_B T}{2\pi \hbar^2}\right)^{3/2} e^{(E_v-\mu)/k_B T}. \tag{6.165}$$

6.4.4 Solution to question 4.3.4

1. Use $n_c = 2\left(\frac{m_c^* k_B T}{2\pi\hbar^2}\right)^{3/2} e^{(\mu - E_c)/k_B T}$ and set $T = 300K$, $m_c^* = m_{transverse}^* = 0.19 m_e$

   or $m_c^* = m_{longitudinal}^* = 0.98 m_e$, and $\mu - E_c \approx -\frac{1}{2} E_g = -\frac{1}{2}(1.11$ eV$)$ and we get

   $n_c = 9.80 \times 10^{14}$ m^{-3} or $n_c = 1.148 \times 10^{16}$ m^{-3}, respectively.

   Note that the conduction band of Si is anisotropic, and thus there are different effective masses in different directions (in reciprocal space).

2.
   - The density of silicon is 2.33 g cm^{-3} and the relative atomic mass of silicon is 28 g mol^{-1}.
   - The volume of 1 mol of silicon $= \frac{28}{2.33} = 12.02$ cm^3.
   - Its intrinsic concentration $= \frac{9.80 \times 10^{14} m^{-3}}{100^3} \times 12.02 = 1.18 \times 10^9$ or 1.38×10^{11}.
   - One phosphorous donor gives one donor electron, 0.01% of 1 mol $= 0.0001$ mol of P gives $0.0001 \times 6.023 \times 10^{23} = 6.023 \times 10^{19}$.
   - The ratio of $\frac{donor}{intrinsic} = \frac{6.023 \times 10^{19}}{1.18 \times 10^9} = 5.11 \times 10^9$ or 4.36×10^8.
   - Comment on the physics: indeed, the donor electrons overwhelm the intrinsic electrons in the CB. This is exactly what technology wants: we control the electronic properties and override the intrinsic properties and then we can make the material do what we want.

3.
   - Use Drude conductivity $\sigma = \frac{n_c e^2 \tau}{m^*}$ and $1.1 = \frac{\sigma'}{\sigma} = \frac{n_c'}{n_c} = \frac{T'^{3/2} e^{(\mu - E_c)/k_B T'}}{T^{3/2} e^{(\mu - E_c)/k_B T}}$.
   - For GaAs, use $\mu - E_c \approx -\frac{1}{2} E_g = -\frac{1}{2}(1.43$ eV$)$.
   - Use $T = 300$ K.
   - Solve the equation $1.1 = \left(\frac{T'}{T}\right)^{3/2} e^{\frac{\mu - E_c}{k_B}\left(\frac{1}{T'} - \frac{1}{T}\right)}$ numerically on a website such as https://www.wolframalpha.com to get $T' = 300.983$ K.
   - Comment on the physics: although the intrinsic conductivity is strongly affected by temperature, in a doped semiconductor, the dopant carriers dominate the effects. Since the dopant carriers do not change much (as they are already excited), the conductivity of the doped semiconductor does not change much with temperature.

6.4.5 Solution to question 4.3.5

1. $E_{donor} = -\frac{1}{\varepsilon_r^2} \frac{m_c^*}{m_e} \times 13.6$ eV $= -\frac{1}{18^2} \times 0.015 \times 13.6$ eV $= 0.000\,63$ eV.

2.
   - The Bohr radius for hydrogen atom $a_0 = \frac{4\pi\varepsilon_0 \hbar^2}{m_e e^2} = 5.29 \times 10^{-11}$ m.

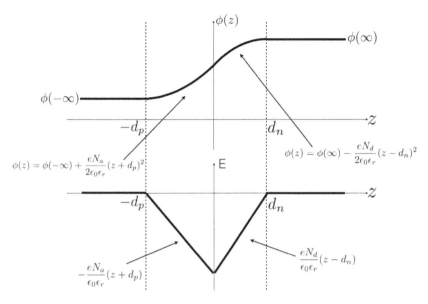

Figure 6.22. For solution 6.4.6: ABOVE: electric potential in a p–n junction. BELOW: the electric field is simply the negative gradient of the electric potential.

- So
$$a_{\text{donor}} = \frac{4\pi\epsilon_0\epsilon_r\hbar^2}{m_c^*e^2} = \frac{\epsilon_r m_e}{m_c^*}\frac{4\pi\epsilon_0\hbar^2}{m_e e^2} = \frac{18}{0.015} \times 5.29 \times 10^{-11} = 6.35 \times 10^{-8}\,\text{m}.$$

3.
- Take the orbit to be spherical; we want the number of spheres in 1 m³.
- Number of spheres in $1\,\text{m}^3 = \left(\frac{1}{2\times 6.35\times 10^{-8}}\right)^3 = 4.88 \times 10^{20}\,\text{m}^{-3}$.
- Comment on the physics: when orbits overlap, the donor electrons can hop between orbits. This implies that the impurity states start to form an impurity band. We are estimating the concentration at which impurity bands start to form.

6.4.6 Solution to question 4.3.6

The electric field is the negative gradient of the electric potential. See figure 6.22.

6.4.7 Solution to question 4.3.7

Since $N_d = N_a$, we have $d_p = d_n = d$. Thus 75Å is $2d$ (see figure 6.23) and using $d = \sqrt{\frac{\epsilon_0\epsilon_r E_g}{e^2 N}}$ with $\epsilon_r = 13$ and $E_g = 1.43$ eV, we have $N = 4.56 \times 10^{24}\,\text{m}^{-3}$.

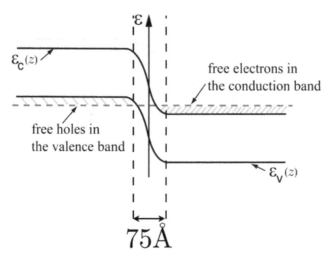

Figure 6.23. An energy diagram of the tunnel diode for solution 6.4.7. Reprinted by permission from Springer Nature [1], Copyright (2018).

6.5 Solutions for exercise 5.5

6.5.1 Solution to question 5.5.1

We use Larmor/Langevin diamagnetism.

$$\chi = -\frac{N}{V}\frac{e^2\mu_0}{6m_e}\sum_{i=1}^{Z}\langle 0|r_i^2|0\rangle$$

| use $\frac{N}{V} = 1 \times 10^{20}$, and there is only one electron, $|0\rangle = |n = 1\rangle$

| in that state, $\langle r^2 \rangle = 3a^2$, $a = 5.29 \times 10^{-11}$ m (Bohrradius) (6.166)

$$\chi = -1 \times 10^{20}\frac{(1.602 \times 10^{-19})^2 \times 4\pi \times 10^{-7}}{6 \times 9.11 \times 10^{-31}} \times 3(5.29 \times 10^{-11})^2 \quad (6.167)$$

$$\chi = -4.95 \times 10^{-15}. \quad (6.168)$$

We then use Curie/Langevin paramagnetism (at 100 K):

$$\chi = \frac{\mu_0 g_{\text{eff}}^2 \mu_B^2 J(J+1)}{3k_B T} \times \frac{N}{V}$$

| for $n = 1$state, $L = 0$, $S = \frac{1}{2} \Rightarrow J = \frac{1}{2}$, $g_{\text{eff}} = 2$ and $\mu_B = 9.27 \times 10^{-24}$ (6.169)

$$\chi = \frac{4\pi \times 10^{-7} \times 4 \times \left(9.27 \times 10^{-24}\right)^2 \frac{1}{2}\left(\frac{1}{2}+1\right)}{3 \times 1.38 \times 10^{-23} \times 100} \times 1 \times 10^{20} \quad (6.170)$$

$$\chi = 7.83 \times 10^{-12}. \tag{6.171}$$

6.5.2 Solution to question 5.5.2

For iron, the lattice constant is $a = 2.87 \times 10^{-10}$ m and since it has a BCC structure, the nearest neighbour (NN) $= \frac{a\sqrt{3}}{2}$. From experimental measurements, the magnetic moment per Fe atom $\approx 2.2\mu_B$.

$$E \approx \frac{\mu_0}{4\pi \left(\frac{a\sqrt{3}}{2} \right)^3} (2.2\mu_B)^2 \tag{6.172}$$

$$E = \frac{4\pi \times 10^{-7}}{4\pi \left(\frac{2.87 \times 10^{-10} \times \sqrt{3}}{2} \right)^3} \times (2.2 \times 9.27 \times 10^{-24})^2 \tag{6.173}$$

$$E = 2.71 \times 10^{-24} \, \text{J} \tag{6.174}$$

The exchange energy is estimated to be $\mathscr{J} \, k_B T_C = 1.38 \times 10^{-23} \times 1043 = 1.44 \times 10^{-20}$ J. So the exchange energy is about 3000 times larger than the dipole energy.

6.5.3 Solution to question 5.5.3

We simply redo the calculation in the lecture notes with the extra term $\vec{S}_N \cdot \vec{S}_{N+1} = \vec{S}_N \cdot \vec{S}_1$ where $N + 1$ is treated as 1.

$$H_{\text{Heisenberg}} = -\mathscr{J} \sum_{i=1}^{N} \vec{S}_i \cdot \vec{S}_{i+1} \tag{6.175}$$

$$H_{\text{Heisenberg}} = -\mathscr{J} \sum_{i=1}^{N} \left(S_i^z S_{i+1}^z + \frac{1}{2} S_i^+ S_{i+1}^- + \frac{1}{2} S_i^- S_{i+1}^+ \right) \tag{6.176}$$

$$H_{\text{Heisenberg}}|0\rangle = -\mathscr{J} \sum_{i=1}^{N} S_i^z S_{i+1}^z |0\rangle \tag{6.177}$$

$$H_{\text{Heisenberg}}|0\rangle = -\mathscr{J} \sum_{i=1}^{N} \left(\frac{1}{2} \right)\left(\frac{1}{2} \right)|0\rangle \tag{6.178}$$

$$H_{\text{Heisenberg}}|0\rangle = -\frac{\mathscr{J}N}{4} \tag{6.179}$$

$$H_{\text{Heisenberg}}|1_i\rangle = -\mathscr{J}\left(\frac{N-2}{4} - \frac{1}{4} - \frac{1}{4}\right)|1_j\rangle - \frac{\mathscr{J}}{2}\left(|i_{j+1}\rangle + |1_{j-1}\rangle\right) \tag{6.180}$$

$$H_{\text{Heisenberg}}|1_i\rangle = -\mathscr{J}\left(\frac{N}{4} - 1\right)|1_j\rangle - \frac{\mathscr{J}}{2}\left(|i_{j+1}\rangle + |1_{j-1}\rangle\right) \tag{6.181}$$

The plane wave state is the same $|q\rangle = \frac{1}{\sqrt{N}}\sum_{j=1}^{N} e^{iqR_j}|1_j\rangle$.

$$H_{\text{Heisenberg}}|q\rangle = \frac{1}{\sqrt{N}}\sum_{j=1}^{N} e^{iqR_j}\left[-\mathscr{J}\left(\frac{N}{4} - 1\right)|1_j\rangle - \frac{\mathscr{J}}{2}\left(|i_{j+1}\rangle + |1_{j-1}\rangle\right)\right]$$

| write $R_j + 1 = R_{j+1}$ and $R_j - a = R_{j-a}$
| due to periodic BC, $R_{N+1} = R_1$, $R_0 = R_N$, $|1_{N+1}\rangle = |1_1\rangle$ and $|1_0\rangle = |1_N\rangle$ $\qquad$ (6.182)
| so $\frac{1}{\sqrt{N}}\sum_{j=1}^{N} e^{iqR_{j\pm1}}|1_{j\pm1}\rangle = \frac{1}{\sqrt{N}}\sum_{j=1}^{N} e^{iqR_j}|1_j\rangle = |q\rangle$

$$H_{\text{Heisenberg}}|q\rangle = \left[-\mathscr{J}\left(\frac{N}{4} - 1\right) - \frac{\mathscr{J}}{2}\left(e^{-iqa} + e^{iqa}\right)\right]|q\rangle \tag{6.183}$$

$$H_{\text{Heisenberg}}|q\rangle = \left[-\frac{JN}{4} + \mathscr{J}(1 - \cos(qa))\right]|q\rangle \tag{6.184}$$

6.5.4 Solution to question 5.5.4

The explicit expression for $H_{\text{Heisenberg}}$ is

$$H_{\text{Heisenberg}} = \left|\mathscr{J}\right|\left(S_1^z S_2^z + S_2^z S_3^z + S_3^z S_4^z + S_4^z S_1^z\right) + \frac{|\mathscr{J}|}{2}\left(S_1^- S_2^+ + S_2^- S_3^+ + S_3^- S_4^+ + S_4^- S_1^+\right)$$
$$+ \frac{|\mathscr{J}|}{2}\left(S_1^+ S_2^- + S_2^+ S_3^- + S_3^+ S_4^- + S_4^+ S_1^-\right) \tag{6.185}$$

Use identities:

$$S_i^z|\uparrow_i\rangle = \frac{1}{2}|\uparrow_i\rangle, \quad S_i^z|\downarrow_i\rangle = -\frac{1}{2}|\downarrow_i\rangle, \quad S_i^+|\downarrow_i\rangle = |\uparrow_i\rangle, \quad S_i^-|\uparrow_i\rangle = |\downarrow_i\rangle. \tag{6.186}$$

The rest is hard work.

$$H_{\text{Heisenberg}}|0\rangle = -|\mathscr{J}||0\rangle$$
$$+ \frac{|\mathscr{J}|}{2}(|\downarrow_1 \uparrow_2 \uparrow_3 \downarrow_4\rangle + |\uparrow_1 \downarrow_2 \downarrow_3 \uparrow_4\rangle + |\uparrow_1 \uparrow_2 \downarrow_3 \downarrow_4\rangle + |\downarrow_1 \downarrow_2 \uparrow_3 \uparrow_4\rangle)) \tag{6.187}$$
$$\Longrightarrow \langle 0|H_{\text{Heisenberg}}|0\rangle = -|\mathscr{J}|$$

$$H_{\text{Heisenberg}}|0'\rangle = -2|\mathscr{J}||\uparrow_1 \downarrow_2 \uparrow_3 \downarrow_4\rangle$$
$$+ |\mathscr{J}|(|\downarrow_1 \uparrow_2 \uparrow_3 \downarrow_4\rangle + |\uparrow_1 \downarrow_2 \downarrow_3 \uparrow_4\rangle + |\uparrow_1 \uparrow_2 \downarrow_3 \downarrow_4\rangle + |\downarrow_1 \downarrow_2 \uparrow_3 \uparrow_4\rangle)$$
$$- \frac{|\mathscr{J}|}{2}(|\downarrow_1 \uparrow_2 \downarrow_3 \uparrow_4\rangle + |\uparrow_1 \downarrow_2 \uparrow_3 \downarrow_4\rangle)$$
$$- \frac{|\mathscr{J}|}{2}(|\downarrow_1 \uparrow_2 \downarrow_3 \uparrow_4\rangle + |\uparrow_1 \downarrow_2 \uparrow_3 \downarrow_4\rangle)$$
$$- 2|\mathscr{J}||\downarrow_1 \uparrow_2 \downarrow_3 \uparrow_4\rangle \tag{6.188}$$
$$+ |\mathscr{J}|(|\downarrow_1 \downarrow_2 \uparrow_3 \uparrow_4\rangle + |\uparrow_1 \uparrow_2 \downarrow_3 \downarrow_4\rangle + |\uparrow_1 \downarrow_2 \downarrow_3 \uparrow_4\rangle + |\downarrow_1 \uparrow_2 \uparrow_3 \downarrow_4\rangle)$$
$$- \frac{|\mathscr{J}|}{2}(\uparrow_1 \downarrow_2 \uparrow_3 \downarrow_4\rangle + |\downarrow_1 \uparrow_2 \downarrow_3 \uparrow_4\rangle)$$
$$- \frac{|\mathscr{J}|}{2}(\uparrow_1 \downarrow_2 \uparrow_3 \downarrow_4\rangle + |\downarrow_1 \uparrow_2 \downarrow_3 \uparrow_4\rangle)$$

$$H_{\text{Heisenberg}}|0'\rangle = -2|\mathscr{J}||0'\rangle \tag{6.189}$$

6.5.5 Solution to question 5.5.

We assume the next nearest neighbour (NNN) Weiss mean field theory of ferrimagnetism.

1. This model includes NNN interactions. There are two effective fields.

$$B_\uparrow = -|\lambda|M_\downarrow + \lambda_\uparrow M_\uparrow + B \tag{6.190}$$

$$B_\downarrow = -|\lambda|M_\uparrow + \lambda_\downarrow M_\downarrow + B \tag{6.191}$$

where

- The term $-|\lambda|M_\downarrow$ refers to the NN mean field 'seen' by up-spins. $\lambda < 0$ and is related to the exchange constant $\mathscr{J}_{\uparrow\downarrow}$. The same explanation goes for the term $-|\lambda|M_\uparrow$. We have also assumed $|\mathscr{J}_{\uparrow\downarrow}| = |\mathscr{J}_{\downarrow\uparrow}|$.
- The term $\lambda_\uparrow M_\uparrow$ refers to the NNN mean field 'seen' by the up-spins. $\lambda_\uparrow > 0$ and is related to the exchange constant $\mathscr{J}_{\uparrow\uparrow}$. The same goes for the term $\lambda_\downarrow M_\downarrow$, where $\lambda_\downarrow > 0$ and is related to $\mathscr{J}_{\downarrow\downarrow}$.

2. The Hamiltonian is

$$H = \sum_{i=1} g_\uparrow \mu_B \vec{J}_{\uparrow i} \cdot \vec{B}_\uparrow + \sum_{i=1} g_\downarrow \mu_B \vec{J}_{\downarrow i} \cdot \vec{B}_\downarrow. \tag{6.192}$$

Again, we can carry over the expression from Curie paramagnetism.

$$M_\uparrow = g_\uparrow \mu_B J_\uparrow B_{J_\uparrow}\left(\frac{g_\uparrow \mu_B B_\uparrow}{k_B T} J_\uparrow\right) \tag{6.193}$$

$$M_\downarrow = g_\downarrow \mu_B J_\downarrow B_{J_\downarrow}\left(\frac{g_\downarrow \mu_B B_\downarrow}{k_B T} J_\downarrow\right) \tag{6.194}$$

3.

   (a) We want the expression for T_C, so put $B \to 0$, then make the small expansion $B_J(xJ) \approx \frac{J+1}{3J}xJ$

$$M_\uparrow \approx g_\uparrow \mu_B \frac{J_\uparrow + 1}{3J_\uparrow} \frac{g_\uparrow \mu_B(-|\lambda|M_\downarrow + \lambda_\uparrow M_\uparrow)}{k_B T} J_\uparrow \tag{6.195}$$

$$M_\uparrow = \frac{g_\uparrow^2 \mu_B^2 J_\uparrow(J_\uparrow + 1)(-|\lambda|M_\downarrow + \lambda_\uparrow M_\uparrow)}{3k_B T}$$

$$\tag{6.196}$$

| define the Curie constant $C_\uparrow = \dfrac{g_\uparrow^2 \mu_B^2 J_\uparrow(J_\uparrow + 1)}{3k_B}$

$$M_\uparrow T = -C_\uparrow |\lambda| M_\downarrow + C_\uparrow \lambda_\uparrow M_\uparrow. \tag{6.197}$$

Similarly, the other equation is

$$M_\downarrow T = -C_\downarrow |\lambda| M_\uparrow + C_\downarrow \lambda_\downarrow M_\downarrow \tag{6.198}$$

$$\Longrightarrow \begin{pmatrix} T - C_\uparrow \lambda_\uparrow & |\lambda| C_\uparrow \\ |\lambda| C_\downarrow & T - C_\downarrow \lambda_\downarrow \end{pmatrix} \begin{pmatrix} M_\uparrow \\ M_\downarrow \end{pmatrix} = 0 \tag{6.199}$$

for non−trivial solutions $\underset{\Longrightarrow}{\det} \begin{vmatrix} T - C_\uparrow \lambda_\uparrow & |\lambda| C_\uparrow \\ |\lambda| C_\downarrow & T - C_\downarrow \lambda_\downarrow \end{vmatrix} = 0 \tag{6.200}$

$$(T - C_\uparrow \lambda_\uparrow)(T - C_\downarrow \lambda_\downarrow) - |\lambda|^2 C_\uparrow C_\downarrow = 0 \tag{6.201}$$

$$T^\pm = \frac{1}{2}\left(C_\uparrow \lambda_\uparrow + C_\downarrow \lambda_\downarrow \pm \sqrt{(C_\uparrow \lambda_\uparrow - C_\downarrow \lambda_\downarrow)^2 + 4|\lambda|^2 C_\downarrow C_\uparrow}\right) \tag{6.202}$$

   (b) The choice is that $T_C = T^+$. This is because if we set $\lambda_\uparrow \to 0$, $\lambda_\downarrow \to 0$, meaning we ignore NNN interactions, we have $T^+ = |\lambda|\sqrt{C_\downarrow C_\uparrow}$, which looks like the Néel temperature T_N. If we assume the two sublattices are the same, $C_\downarrow = C_\uparrow$, then we get $T^+ = T_N$ exactly.

$$T_C = T^+ = \frac{1}{2}\left(C_\uparrow \lambda_\uparrow + C_\downarrow \lambda_\downarrow + \sqrt{(C_\uparrow \lambda_\uparrow - C_\downarrow \lambda_\downarrow)^2 + 4|\lambda|^2 C_\downarrow C_\uparrow}\right) \tag{6.203}$$

4. Next, we address the susceptibility for $T > T_C$; we put back the external field B and make the small expansion $B_J(xJ) \approx \frac{J+1}{3J}xJ$:

$$M_\uparrow T = C_\uparrow(B - |\lambda|M_\downarrow + \lambda_\uparrow M_\uparrow) \tag{6.204}$$

$$M_\downarrow T = C_\downarrow(B - |\lambda|M_\downarrow + \lambda_\uparrow M_\uparrow) \tag{6.205}$$

$$\Longrightarrow \begin{pmatrix} T - C_\uparrow \lambda_\uparrow & |\lambda| C_\uparrow \\ |\lambda| C_\downarrow & T - C_\downarrow \lambda_\downarrow \end{pmatrix} \begin{pmatrix} M_\uparrow \\ M_\downarrow \end{pmatrix} = B \begin{pmatrix} C_\uparrow \\ C_\downarrow \end{pmatrix}. \tag{6.206}$$

Invert the matrix and solve $M_\uparrow$ and $M_\downarrow$.

$$\begin{pmatrix} M_\uparrow \\ M_\downarrow \end{pmatrix} = \frac{B}{\det(\cdots)} \begin{pmatrix} T - C_\downarrow \lambda_\downarrow & -|\lambda| C_\uparrow \\ -|\lambda| C_\downarrow & T - C_\uparrow \lambda_\uparrow \end{pmatrix} \begin{pmatrix} C_\uparrow \\ C_\downarrow \end{pmatrix} \tag{6.207}$$

| write $\det(\cdots) = (T - C_\uparrow \lambda_\uparrow)(T - C_\downarrow \lambda_\downarrow) - |\lambda|^2 C_\uparrow C_\downarrow = (T - T^+)(T - T^-)$

$$\begin{pmatrix} M_\uparrow \\ M_\downarrow \end{pmatrix} = \frac{B}{(T - T^+)(T - T^-)} \begin{pmatrix} C_\uparrow T - C_\uparrow C_\downarrow \lambda_\downarrow - |\lambda| C_\uparrow C_\downarrow \\ C_\downarrow T - C_\uparrow C_\downarrow \lambda_\uparrow - |\lambda| C_\uparrow C_\downarrow \end{pmatrix} \tag{6.208}$$

$$\chi = \frac{\mu_0 M}{B} = \frac{\mu_0 (M_\uparrow + M_\downarrow)}{B}$$

$$\chi = \frac{\mu_0}{(T - T^+)(T - T^-)} (T(C_\uparrow + C_\downarrow) - C_\uparrow C_\downarrow (\lambda_\downarrow + \lambda_\uparrow + 2|\lambda|)) \tag{6.209}$$

5. We can now treat some special cases:
   (a) Ferrimagnetism with the NN mean field.
      In this case, we set $\lambda_\downarrow = 0 = \lambda_\uparrow$.

$$T_C = T^+ = |\lambda| \sqrt{C_\downarrow C_\uparrow} \tag{6.210}$$

For $T > T_C$, the susceptibility is

$$\chi = \frac{\mu_0}{T^2 - |\lambda|^2 C_\uparrow C_\downarrow} (T(C_\uparrow + C_\downarrow) - 2|\lambda| C_\uparrow C_\downarrow) \tag{6.211}$$

$$\chi = \frac{\mu_0}{T^2 - T_C^2} (T(C_\uparrow + C_\downarrow) - 2|\lambda| C_\uparrow C_\downarrow). \tag{6.212}$$

   (b) Antiferromagnetism with the NNN mean field.
      This means the two sublattices are identical, so we set $C_\uparrow = C_\downarrow = C$ and $\lambda_\uparrow = \lambda_\downarrow = \lambda' > 0$.

$$T_C = T^+ = \frac{1}{2} \left(C\lambda' + C\lambda' + \sqrt{(C\lambda' - C\lambda')^2 + 4C^2 |\lambda|^2} \right) \tag{6.213}$$

$$T_C = C(\lambda' + |\lambda|) \tag{6.214}$$

$$T_C = T_N \text{ (which is the Néel temperature withthe NNN mean field)} \tag{6.215}$$

For $T > T_C$, the susceptibility is

$$\chi = \frac{\mu_0}{(T - T^+)(T - T^-)} (T(C + C) - C^2 (2\lambda' + 2|\lambda|)) \tag{6.216}$$

| note that $T^- = C(\lambda' - |\lambda|)$

$$\chi = \frac{2C\mu_0}{T - C(\lambda' - |\lambda|)} \tag{6.217}$$

| if we define $T_N' = C(|\lambda| - \lambda')$

$$\chi = 2C\mu_0 \frac{1}{T + T_N'} \tag{6.218}$$

It is interesting to note that $\frac{T_N}{T_N'} = \frac{|\lambda| + \lambda'}{|\lambda| - \lambda'}$, and that the 'special' temperature in χ is not the critical temperature.

(c) Antiferromagnetism with the NN mean field.

We simply set $\lambda' = 0$.

$$T_C = T^+ = C|\lambda| = \frac{g_{\text{eff}}^2 \mu_B^2 J(J + 1)}{3k_B} |\lambda| = T_N \tag{6.219}$$

$$\chi = 2C\mu_0 \frac{1}{T + C|\lambda|} \tag{6.220}$$

| from above, $T_N = C|\lambda|$

$$\chi = \frac{2\mu_0 T_N}{|\lambda|} \frac{1}{T + T_N} \tag{6.221}$$

which are exactly the same results as in the lecture notes. For this case, the 'special' temperature in χ is the critical temperature.

(d) Ferromagnetism with the NN mean field.

We simply set $|\lambda| = 0$.

$$T_C = T^+ = C\lambda' = \frac{g_{\text{eff}}^2 \mu_B^2 J(J + 1)}{3k_B} \lambda' \tag{6.222}$$

$$\chi = 2C\mu_0 \frac{1}{T - C\lambda'} = \frac{2\mu_0 T_C}{\lambda'} \frac{1}{T - T_C} \tag{6.223}$$

The T_C is the same as in the ferromagnetic case except for the notation of λ'. The susceptibility has an extra factor of two because in the ferrimagnetic model we started with, we had two sublattices. Therefore, we should divide by two to match the ferromagnetic model. Indeed, we do get exactly the same expressions as in the ferromagnetic case.

Reference

[1] Quinn J J and Yi K-S 2018 *Solid State Physics: Principles and Applications* 2nd edn (Springer)

Printed in the USA
CPSIA information can be obtained
at www.ICGtesting.com
JSHW061315120124
55268JS00004B/36